WIDE-RANGE AND MULTI-FREQUENCY ANTENNAS

Boris Levin
Holon Institute of
Technology
Lod, Israel

CRC Press
Taylor & Francis Group
Boca Raton London New York

CRC Press is an imprint of the
Taylor & Francis Group, an **informa** business
A SCIENCE PUBLISHERS BOOK

CRC Press
Taylor & Francis Group
6000 Broken Sound Parkway NW, Suite 300
Boca Raton, FL 33487-2742

First issued in paperback 2021

© 2019 by Taylor & Francis Group, LLC
CRC Press is an imprint of Taylor & Francis Group, an Informa business

No claim to original U.S. Government works

Version Date: 20180921

ISBN-13: 978-0-367-78023-4 (pbk)
ISBN-13: 978-1-138-48744-4 (hbk)

Library of Congress Cataloging-in-Publication Data

Names: Levin, Boris (Electrical engineer), author.
Title: Wide-range and multi-frequency antennas / Boris Levin, Holon Institute of Technology, Lod, Israel.
Description: Boca Raton, FL : CRC Press/Taylor & Francis Group, [2018] | "A Science Publishers book." | Includes bibliographical references and index.
Identifiers: LCCN 2018041506 | ISBN 9781138487444 (hardback)
Subjects: LCSH: Ultra-wideband antennas.
Classification: LCC TK7871.67.U45 L48 2018 | DDC 621.3841/35--dc23
LC record available at https://lccn.loc.gov/2018041506

Visit the Taylor & Francis Web site at
http://www.taylorandfrancis.com

and the CRC Press Web site at
http://www.crcpress.com

Contents

Introduction

The book is devoted to the development of wide-range and multi-frequency antennas. Both antenna families allow expansion of the capabilities of structures providing transmission and reception of electromagnetic signals; in other words, the functioning of radio communication channels. The proposed solutions of the problem are based on using known and developing new methods of analyzing and synthesizing antenna devices.

Analysis of antennas is a so-called direct problem whose solution allows us to determine the properties of antennas and calculate their electrical characteristics. Synthesis, or inverse problem, makes it possible to create devices with high characteristics by developing new antenna variants, using new materials and new technologies, or by optimizing antennas' dimensions and magnitudes of elements employed in a device. As a rule, a successful solution to both the tasks is based on the development of new theoretical methods and new designs.

The task of expanding the frequency range involves creation of a radiator operating in a wide frequency range and providing high electrical characteristics at these frequencies, including the required level of efficiency and matching with the cable (generator), as well as the desired shape of the radiation pattern. For a long time, engineers tried solving this problem by finding the law of current distribution along the radiator, which provides the required characteristics. At the same time, the problem of creating this hypothetical current distribution was not raised; the question of optimizing characteristics by choosing antenna dimensions and magnitudes of antenna loads remained open. The task of optimizing characteristics by means of choosing antenna dimensions by methods of mathematical programming was first put in relation to the Yagi-Uda antenna. Her decision confirmed the correctness of the chosen approach.

Subsequently, the issue of creating a broadband radiator was raised. This task was solved by applying concentrated capacitive loads (capacitors) included in a radiator wire and forming an in-phase current with a predetermined law of its distribution along the antenna. The general methodology of calculating antennas with distributed surface impedance and concentrated loads is considered in Chapter 1. The solution is based on an understanding of the benefit of in-phase current distribution and Hallen's hypothesis on the utility of capacitive loads, whose magnitudes vary along the radiator axis in accordance with an exponential law.

The chosen approach confirmed Hallen's hypothesis and demonstrated the effectiveness of a proposed approximate methods of calculating capacitive loads, viz. a method of a long line with distributed surface impedance and a method of a metallic long line with concentrated loads. These methods are described in Chapter 2. Further, the loads calculated by these methods were used as initial magnitudes in the process of numerical solving of the problem by methods of mathematical programming. The calculations results showed that these methods allow to obtain the required level of antenna matching with a cable in a frequency ratio of about 10. The efficiency of the created methodology is obvious in comparison to the trial-and-error method that uses an exact calculation or an approximate measurement of randomly selected radiators.

The used method helped to solve three additional tasks. The first task was to select the load magnitudes that provide a given current distribution in the necessary frequency range. Thus, in particular, the efforts aimed at finding the distribution of a current providing the required electrical characteristics were justified. The second result was the choice of loads, which made it possible to reduce the distortion of the antenna directional patterns by closely located superstructures. Finally, this method was used to select the loads included in the wires of the *V*-antenna that in the wide frequency range help to produce a high directivity along a bisector of an angle between the antenna wires. These tasks are discussed in Chapter 3.

In the course of the work, a reasonable sequence of solving each problem was determined. At the first stage, one needs to develop an approximate method of analysis, whose results are used as the initial values for the numerical solution of the problem by methods of mathematical programming.

Since the placement of concentrated capacitive loads along the radiator allows achievement of not only a high level of antenna matching with the cable, but also the required directional patterns in a wide frequency range, it is advisable to consider the application results of these loads in directional antennas.

For the analysis of directional characteristics of antennas, the author proposes rigorous theoretical methods—a method for calculating the directional patterns of radiators with a given current distribution and a method of electrostatic analogy for calculating fields of complex radiating structures comprising many elements. The first method helps to obtain in the wide frequency range by means of antennas with capacitive loads the required shape of the directional pattern providing a significant increase in the communication distance. To evaluate this directional pattern, a special parameter was introduced, called the pattern factor.

Using the second method, the optimal electrical characteristics of director and log-periodic antennas with concentrated capacitive loads were obtained. These results make it possible to increase the frequency range of director antennas and to develop proposals that allow decrease dimensions of log-periodic antennas. New methods and results of their application are detailed in Chapter 4.

Capacitive loads can significantly expand the working range of linear antennas. A similar problem can be solved for complementary and self-complementary antennas. These antennas are based on the principle of duality and consist of an equal number of metal (electric) and slot (magnetic) radiators. Mutual replacement of these radiators creates fields with an identical structure. As is known, the uniqueness of these structures is that the infinite structures possess a special property—a high level of matching with the cable and this property is partially preserved, if the dimensions of the structures are finite.

A first step forward compared with a simple self-complementary antenna in the form of a flat metal dipole with an angular width equal to an angular width of the slot, was the creation of a flat self-complementary antenna with rotation symmetry consisting of several metal and slot radiators. The antennas with rotation symmetry allow use of different variants in connecting the generator poles (cable leads-in) to antenna poles in order to improve their mutual matching. Another step forward is the development of three-dimensional (volumetric) antennas. Their appearance became possible when it was proved that the antenna based on the principle of complementarity can be located not only on the plane, but also on the surface of rotation; for example, on the surface of a circular cone or paraboloid. Characteristics of simple complementary antennas consisting of one metal and one-slot dipole and located on the surface of the horizontal circular cone and paraboloid, as well as on the faces of a horizontally-lying regular and irregular pyramid, are discussed in Chapter 5.

Multi-element self-complementary antennas with rotation symmetry, both flat and volumetric, and consisting of a few metal and slot radiators can be applied, using different variants of connections with poles of generator and with leads-in of cable. The characteristics of these antennas in dependence from the number of dipoles and variants of connecting are described in Chapter 6. They allow decrease in the wave impedance of the antenna to the magnitude of the wave impedance of a standard cable.

In the second part of the book multi-frequency antennas are considered. Their frequency range is not continuous, but such antennas in a wide range can create a large number of operating points, i.e., they provide antenna characteristics that are close to those of wide-band antennas. The analysis of such antennas is based on the method proposed by Pistolkors for the calculation of wire structures consisting of long metal wires parallel to each other and to a horizontal metal plane (to the ground). This method and example of its application to an antenna with meandering load are given in Chapter 7. The method ensures the calculation of currents and voltages in the structure.

Placing these structures vertically and perpendicular to the ground allows us to analyze folded vertical antennas, symmetrical and asymmetric, including wires of different lengths and diameters. The generalization of the calculating method for arbitrary wires with different propagation constants allows us to consider structures of different wires, for example, a folded impedance antenna consisting of a metal wire and a wire with a magneto-dielectric shell (Chapter 8).

Multi-folded structures without increasing the antenna height sharply increase the number of points of a series resonance in a given frequency range and provide a freedom of the points movement within the interval. There was calculated a loss resistance in these structures caused by different reasons (Chapter 9). Chapter 10 gives a brief description of multi-wire antennas, including cage antennas. A detailed description is given of one of the multi-radiator antenna that has found practical application due to the inclusion of a complex load in the central (longer) radiator in the form of a parallel connection of a resistor and an inductor.

Of particular interest is the two-tier antenna consisting of two radiators, located one above the other with exciting emfs that provide radiation along the earth's surface. Switching of emfs changes the radiating length of the antenna to provide the desired shape of a radiation pattern in two frequency ranges (Chapter 11). The second section of this Chapter is devoted to asymmetrical coaxial log-periodic antenna, in which a coaxial distribution line replaces the two-wire line and monopoles replace dipoles that allows decrease in the antenna dimensions.

Chapter 12 deals with specific issues. Firstly, this is an actual topic related to the consequences of an emergency appearance of an arbitrary load in an antenna and the effect of this appearance on the characteristics of antennas with different current distributions. The second section presents a rigorous method of analyzing a transparent antenna, taking into account the losses in the antenna film. The solution of the integral equation for the current in the antenna shows that the current along the antenna is distributed according to the sinusoidal law, and the amplitude of the sinusoid decreases exponentially, i.e., the length of the radiating section is substantially shorter than the length of the antenna and, in the first approximation, depends weakly on this length. It is proposed that the antenna variant ensures a high matching level with a cable owing to the use of self-complementary structure with a rotation symmetry. The third section is devoted to calculating the fields of a rectangular loop.

The proposed book is a natural addition to the known monographs. It is intended for professionals, who are engaged in the development, placement and exploitation of antennas and also for lecturers, teachers, students, advisors, etc. The contents of the book can be used for university courses.

At the end of the introduction, I consider that it is necessary to mark one anniversary. Fifty years ago, King's article 'The Linear Antenna – Eighty Years of Progress' was published in *Proceedings IEEE* (Vol. 55, No. 1, 1967). In this article, King reminds us that a hundred years have passed since Maxwell formulated his famous equations and 80 years since Hertz experimentally proved the existence of the wave phenomena predicted by these equations.

This book tells us about the development of one field of antenna engineering over the next half century. This development is described by a man from the Soviet Union who currently resides in Israel. Therefore, it is not surprising that I have never seen the profoundly respected Prof. King. An article dedicated to this issue was written for the anniversary of King's article, but unfortunately it was not published. Therefore, the author decided to present a more comprehensive material to publish in book form.

The proposed book describes the history of creating radiators that provide in a wide frequency range a high level of matching with the cable and a necessary directional pattern that allows increase in the communication distance. High electrical characteristics of the radiator were obtained as a result of including concentrated capacitive loads in its central wire and creating in-phase current, the amplitude of which varies in accordance with the linear or exponential law. Methods of solving this problem and the obtained results are described.

Antenna engineering in the Soviet Union has never lagged behind the world level; suffice it to recall the names of Klyatskin, Pistolkors, Eisenberg, Weinstein, Braude, Drabkin, Shifrin. In 1944 Leontovich and Levin published an article 'On the theory of oscillations excitation in the linear radiators' in the magazine *Journal of Technical Physics*. This article presents the most rigorous integral equation for a current in a thin linear radiator. Ten years later, in the same magazine, Leontovich's disciple Miller published the article entitled 'Application of Uniform Boundary Conditions in the Theory of Thin Antennas'. This article started development of the field of antenna engineering, which is described in the present book.

PART 1
WIDE-RANGE ANTENNAS

Radiators with Distributed Loads

1.1 Radiators with non-zero (impedance) boundary conditions— Constant surface impedance

As stated in the Introduction, applying concentrated capacitive loads (capacitors) played a fundamental role in creating wide-band antenna. Such an antenna is a radiator with loads which are installed along an axis of a linear antenna and form an in-phase current with a given amplitude distribution law. An antenna with surface impedance in the form of a metal rod with a ferrite shell was a first studied radiator of this type [1]. It is the radiator with distributed loads whose study showed that using a ferrite shell allows to change in the range of antenna operating frequencies.

Accordingly, as a first step, it is expedient to consider radiators with distributed loads. From the foregoing it follows that the antenna in the form of a metal rod covered by a layer of magneto dielectric (Fig. 1.1) can serve as an example of such a radiator. Let the rod length be L, the rod radius be a_1 and the outer radius of the shell be a. The boundary conditions on the shell surface have a form

$$\frac{E_z(a,z)+K(z)}{H_\varphi(a,z)}\Big|_{-L\leq z\leq L}=Z(z), \tag{1.1}$$

where $E_z(a, z)$ is the vertical component of an electric field, $H_\varphi(a, z)$ is the azimuthal component of a magnetic field on the antenna surface, $K(z)$ is the exciting (external) electromotive force (emf) and $Z(z)$ is the surface impedance, which in general case depends on coordinate z (it is assumed that the antenna axis coincides with z axis of a cylindrical coordinate system—see Fig. 1.1). The boundary conditions of this type are valid if the

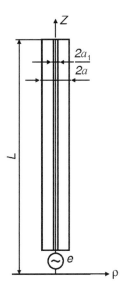

Fig. 1.1: Antenna with distributed surface impedance.

field structure in one of the media (for example, in the magneto-dielectric shell) is known and does not depend on the structure of the field in another medium (in surrounding space). Radiators, on whose surface the boundary conditions (1.1) are fulfilled and the surface impedance changes substantially the current distribution along the antenna, are called radiators with non-zero (impedance) boundary conditions, or simply impedance radiators (it means in the first approximation).

In accordance with the equivalence theorem, when considering electromagnetic fields in a free space surrounding a radiator, an antenna should be replaced by a field at the boundaries. Although in this case, it is possible to operate only with fields, but for clarity and simplicity of reasoning, it is expedient mentally to put a metal coat on the antenna surface. The surface density j_s of a total linear electric current $J(z)$ along the metal coat is related to the strength H of the magnetic field by an expression

$$[e_\rho, H] = j_s,\tag{1.2}$$

where e_ρ is a unit vector directed along the axis ρ. Then

$$H_\varphi(a, z) = j_z(z) = J(z)/(2\pi a).\tag{1.3}$$

Substituting in (1.1) the magnetic field $H_\varphi(a, z)$ and the tangential component of the electric field, we obtain an equation for the current in an impedance radiator, generalizing Leontovich's integral equation for the current in a metal radiator [2]:

$$\frac{d^2 J(z)}{dz^2} + k^2 J(z) = -4\pi j \omega \varepsilon_0 \chi \left[K(z) + W(J,z) - \frac{J(z)Z(z)}{2\pi a} \right], \qquad (1.4)$$

where ω is a circular frequency of a signal, ε_0 is the absolute permittivity of the air, χ is the small parameter and W is a certain functional. This equation must satisfy the condition of no current at the antenna ends. The right side of this equation contains three terms in square brackets: the first term takes into account the exciting emf, the second one—radiation and the third one—the presence of a distributed load.

The equation solution is sought as an expansion into a series in powers of the small parameter χ:

$$J(z) = J_0(z) + \chi J_1(z) + \chi^2 J_2(z) + \dots \qquad (1.5)$$

Usually this parameter is selected in the form $\chi = 1/[2\ln(2L/a)]$ that corresponds to a straight radiator, whose length does not exceed the wavelength. For the no-resonant radiator, when $J_0(z) = 0$, we obtain the following system of equations:

$$\frac{d^2 J_1(z)}{dz^2} + k_1^2 J_1(z) = -4\pi j \omega \varepsilon_0 K(z), \qquad J_1(\pm L) = 0,$$

$$\frac{d^2 J_n(z)}{dz^2} + k_1^2 J_n(z) = -4\pi j \omega \varepsilon_0 W(J_{n-1}, z), J_n(\pm L) = 0, n > 1, \qquad (1.6)$$

where in the case of constant impedance

$$k_1^2 = const = k^2 - U = k^2 - j2\omega\varepsilon_0 \chi Z/a = k^2 - j2k\chi Z/(aZ_0), \qquad (1.7)$$

$Z_0 = 120\pi$ is the wave impedance of a free space.

If the quantities k^2 and U have the same order of smallness, then the surface impedance essentially affects the current distribution. One should attribute to the magnitude $k_1 = \sqrt{k^2 - j2k\chi Z/(aZ_0)}$ the meaning of a new wave propagation constant along an antenna. The ratio k_1/k is usually named slowing. From the first equation of the system (1.6), it follows that the current distribution along the antenna has in a first approximation a sinusoidal character:

$$\chi J_1(z) = j\chi \frac{ke}{60 k_1 \cos k_1 L} \sin k_1(L - |z|). \qquad (1.8)$$

The input impedance in that approximation is purely reactive:

$$Z_A = \frac{e}{\chi J_1(0)} = -j\frac{60 k_1}{k\chi} \cot kL. \qquad (1.9)$$

This impedance is equal to the input impedance of a uniform long line, open at the end, in which, wave propagation velocity is k_1/k times smaller than in the air. The propagation constant and the wave impedance of this line are k_1/k times greater than in a metallic long line with identical dimensions:

$$W_1 = 60k_1/(k\chi) = 120\,k_1/k\,\ln\,(2L/a). \tag{1.10}$$

Such a line is called an impedance line. Just as an ordinary long line is an equivalent of a metal radiator, an impedance line is an equivalent of an impedance radiator. If the surface impedance is constant along the antenna length, an equivalent of an impedance radiator is a uniform long line. More details are given in Section 1.2.

In Fig. 1.1 is given an example of an antenna with distributed surface impedance. The antenna is a metal rod of a radius a_1 located along the antenna axis and surrounded by a ferrite shell of a radius a with absolute permeability μ and absolute permittivity ε. The field structure inside the antenna will not change, if, to cut the ferrite shell into parts in the shape of rings, between which infinitely thin and perfectly conductive equidistant metal disks are placed, connected to the metal rod.

If the distances between the disks are small in comparison with the wavelength in ferrite, the antenna structure can be represented as a set of radial lines shorted by the metal rod. By calculating an input impedance Z_{rl} of a radial line, one can find the surface impedance of an antenna. A voltage and a current of the radial line are equal to $U(\rho) = AJ_0(m\rho) + BY_0(m\rho)$ and $J(\rho) = j\dfrac{1}{W_p}[AJ_1(m\rho) + BY_1(m\rho)]$ respectively. Here ρ is a line length,

$W_p = 120\,\pi\,\sqrt{\mu_r/\varepsilon_r}\;h/(2\pi a)$ is a wave impedance of the line, $J_0,\,Y_0,\,J_1$ and Y_1 are Bessel functions, m is the propagation constant in ferrite, μ_r and ε_r are the relative permeability and permittivity of ferrite and h is the disk thickness. Taking into account that in an antenna center $U|_{\rho=a_1} = 0$, we find:

$$Z_p = \left.\frac{U(\rho)}{J(\rho)}\right|_{\rho=a} = -jW_p\,\frac{Y_0(ma)J_0(ma_1)-J_0(ma)Y_0(ma_1)}{Y_1(ma)J_0(ma_1)-J_1(ma)Y_0(ma_1)}. \tag{1.11}$$

Since $U(a) = hE_z(a)$ and $J(a) = 2\pi a H_\varphi(a)$, the surface impedance is

$$Z = \left.\frac{E_z}{H_\varphi}\right|_{\rho=a} = Z_p\,\frac{2\pi a}{h}. \tag{1.12}$$

An analogous result can be obtained by considering the diffraction of a converging cylindrical wave on a circular infinitely long cylinder.

If the antenna diameter is small in comparison with the wavelength in ferrite, i.e., if $ma \ll 1$,

$$Z = j120\pi\mu_r ka \ln(a/a_1),\qquad(1.13)$$

and in accordance with (1.7).

$$k_1 = k\sqrt{1+\mu_r \ln(a/a_1)/\ln(2L/a)}\qquad(1.14)$$

As it follows from (1.13), the surface impedance of thin antennas is directly proportional to the quantity μ_r and does not depend on ε_r. Thus, to ensure a significant slowing in a thin shell, it is necessary to choose a magneto-dielectric with a high μ_r (ferrite). At first, hoping to achieve high slowing, researchers tried to use dielectric materials. As a rule, they were deeply disappointed at the results obtained. As for the propagation constant of the wave along the antenna in the form of a metal rod with a ferrite shell, it is always greater than in the air.

It is necessary to emphasize that the main result of slowing increase is a decrease of resonant frequencies of an antenna that permits hope for decrease in its length. This result was obtained by calculation and confirmed by an experiment. To a large extent, these results have stimulated the work on expanding the frequency range of antennas.

In Fig. 1.2 the calculated curves for the active and reactive components of the input impedances and the radiation resistances of three asymmetrical radiators are given. Their parameters are summarized in Table 1.1. The calculated radiation resistance can be determined by subtracting firstly, the usual losses from the active component of the input impedance and secondly, the additional losses in a ferrite shell. The additional losses can be determined by introducing a complex magnetic permeability $\mu_r = \mu_1 - j\mu_2$ of ferrite, the second term of which leads to losses. Calculations were made in the range of decameter waves for antennas with a height of 1–2 m with the slowing from 2.3 to 5. The antenna efficiency depends on the frequency, increases with its increase and is average for variant 1 by about 50 per cent and for variant 2 by 80 per cent.

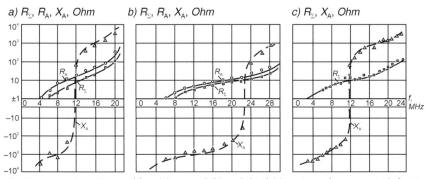

Fig. 1.2: Input impedance of first (a), second (b) and third (c) variant of antenna with ferrite shell.

The experimental method for determining the radiation resistance is based on measuring the signals received by investigated and reference antenna from the same source. As a reference antenna, a linear metal radiator of equal height is used. The ratio of the effective lengths of both antennas is determined by the ratio of the received signals. If the effective length of the reference antenna is known, it is not difficult to find the effective length and radiation resistance of the investigated antenna. This method has high accuracy and allows division of the active component of the input impedance by the radiation resistance and loss resistance.

The experimental results are given in Fig. 1.2 by circles, squares and triangles. They are quite close to the calculation results. Some discrepancy at high frequencies is due to an increase in the magnitude μ_2, which was assumed to be constant in the calculations.

The method of calculating input impedance is described further.

Table 1.1: Parameters of Asymmetrical Radiators.

Variant	L (м)	a(м)	a_1(м)	μ_r	ε_r	k_1/k
1	2.0	0.021	0.007	40	10	3.0
2	1.5	0.15	0.007	25	10	2.3
3	1.0	0.021	0.007	100	10	5.0

1.2 Impedance long line as an approximate equivalent of an impedance radiator

Unlike a metal antenna, the tangential component of the electric field $E_z(a, z)$ on the surface of an impedance radiator is not equal to zero, thus leading to an additional voltage drop on each element dz. Using the boundary condition (1.1) and taking (1.3) into account, we obtain for a line equivalent to a symmetrical impedance radiator

$$dU = ZJ(z)dz/(2\pi a).$$

Thus, an infinitesimal element dz of this line contains in addition to an inductance $d\Lambda = \Lambda_1\, dz$ and a capacitance $dC = C_1 dz$ also an impedance $Zdz/(\pi a)$. A long line, which is equivalent to the impedance antenna, is shown in Fig. 1.3. Here C_1 u Λ_1 are a capacitance and an inductance per unit length of the line. Coefficient 2 takes into account that the radiator consists of two wires. Telegraph equations for such a line:

$$-\frac{dU}{dz} = J(z)\left(j\omega\Lambda_1 + \frac{Z}{\pi a} \right), -\frac{dJ(z)}{dz} = j\omega C_1 U(z), \tag{1.15}$$

from which

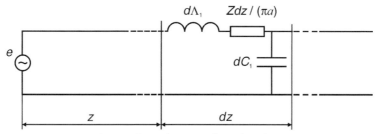

Fig. 1.3: Equivalent impedance long line.

$$\frac{d^2 J(z)}{dz^2} + k_1^2 J(z) = 0, \frac{d^2 U(z)}{dz^2} + k_1^2 U(z) = 0, \tag{1.16}$$

where $k_1^2 = k^2 - jZ\omega C_1/(\pi a)$.

We will consider that the capacitance C_w per unit length of the antenna wire as being equal to the intrinsic capacity of an infinitely long wire of radius a. Since the radius a is much smaller than the length $2L$ of the antenna, the surface of zero potential can be located at a distance $2L$ from the antenna. Then the capacitance C_1 per unit length of the line, which is equivalent to the symmetric impedance antenna, is equal to

$$C_1 = Cw/2 = \pi\varepsilon/\ln(2L/a)$$

and $k_1^2 = k^2 - U = k^2 - j2\omega\varepsilon_0\chi Z/a$, that coincides with (1.7).

Solving equations (1.16) in the usual way, we find the current $J_1(z)$ and the input impedance Z_1 of the impedance line, open at the end:

$$J_1(z) = J_1(0)\frac{\sin k_1(L-z)}{\sin k_1 L}, Z_1 = -jW_1 \cot k_1 L, \tag{1.17}$$

where $J_1(0)$ is the generator current and W_1 is the wave impedance of the long line. The expressions that are obtained are similar to the expressions given for the same quantities in the previous section. This means that in calculating the input impedance of an antenna with constant surface impedance, one can replace this antenna in the first approximation by an equivalent impedance line, which differs from a normal line by the presence of the surface impedance $Z/(\pi a)$ per unit length of the line.

In this case, the antenna current is distributed according to the sinusoidal law with propagation constant k_1, different from the propagation constant k in the metal antenna, and the wave impedance of the impedance radiator is k_1/k times greater than the wave impedance of the metal antenna. Calculation of an equivalent impedance line yields results that coincide in the first approximation with the results of solving the integral equation.

The generalization of the expressions for the current and the input impedance of a metal antenna to the case of an antenna with a constant surface impedance allows application of the induced emf method to calculate an input impedance of an impedance antenna. As is known, this method is based on equating a power created by an emf source (by a generator) and a power radiated by an antenna. In the original version (it is called the first formulation of the method) the complex powers are equated. That causes errors in a presence of losses in the antenna or its environment.

The reason for these errors is the use of a concept of a complex power, consisting of active power and reactive power, since reactive power has no physical meaning. Replacement of a reactive power by an oscillating power [3], i.e., equating of the instantaneous powers flowing from one volume to another, corrects the error, leading to a coincidence of the results of using this method with the results of solving integral equations. This version of the method is called the second formulation of the method of induced emf. Further, speaking of this method we will always bear in mind the second formulation, unless otherwise specified.

The method of induced emf permits derivation of an expression for the input impedance of the radiator, identical to the solution of Leontovich's integral equation in the second approximation. The radiated power is defined as an integral from Poynting vector (from its normal component $S_\rho = - E_z H_\varphi$) along the side surface of an antenna. For the impedance antenna in this expression, the value E_z should be replaced by the difference $E_{z_1} = E_z - H_\varphi Z$, i.e., in accordance with the boundary condition (1.1), it is necessary to subtract from the vertical component E_z of an electric field the voltage drop $H_\varphi Z$ at the antenna itself, since this voltage drop does not make contribution to the antenna radiation. Then for the input impedance of the antenna, we get

$$Z_A = -\frac{1}{J^2(0)} \int_{-L}^{L} \left[E_\varsigma(J) - \frac{ZJ(\varsigma)}{2\pi a} \right] J(\varsigma) d\varsigma, \tag{1.18}$$

and the character of current distribution along the antenna is determined by means of an impedance long line, i.e., in accordance with (1.17).

In a cylindrical coordinate system, using the known expression

$$E_z(\rho, z) = -\frac{j\omega}{k^2} \left(k^2 A_z + \frac{\partial^2 A_z}{\partial z^2} \right) \tag{1.19}$$

and keeping $\rho = a$, we find the longitudinal tangential component of the electric field on the surface of the impedance antenna

$$E_z = -j \frac{30J(0)}{\sin k_1 L} \left[\frac{k^2 - k_1^2}{k} \int_0^L \left(\frac{e^{-jkR}}{R} + \frac{e^{-jkR_+}}{R_+} \right) \sin k_1(L - \varsigma) d\varsigma + \frac{k_1}{k} \left(\frac{e^{-jkR_1}}{R_1} + \frac{e^{-jkR_2}}{R_2} - 2\frac{e^{-jkR_0}}{R_0} \cos k_1 L \right) \right],$$

where $R = \sqrt{(z-\varsigma)^2 + a^2}, R_+ = \sqrt{(z+\varsigma)^2 + a^2}, R_1 = \sqrt{(L-z)^2 + a^2}, R_2 = \sqrt{(L+z)^2 + a^2},$
$R_0 = \sqrt{z^2 + a^2}.$

The substitution of this expression in (1.18) yields

$$Z_A = j\frac{60}{\sin^2 k_1 L}\int_0^L\left[\frac{k^2 - k_1^2}{k}\int_0^L\left(\frac{e^{-jkR}}{R} + \frac{e^{-kR_+}}{R_+}\right)\sin k_1(L-\varsigma)d\varsigma + \frac{k_1}{k}\left(\frac{e^{-jkR_1}}{R_1} + \frac{e^{-jkR_2}}{R_2} - 2\cos k_1 L\frac{e^{-jkR_0}}{R_0}\right)\right]*$$

$$\sin k_1(L-z)dz + \frac{Z}{4\pi a \sin^2 k_1 L}\left(2L - \frac{\sin 2k_1 L}{k_1}\right). \tag{1.20}$$

This expression differs from the analogous expression for a metallic radiator by the presence of an integral term with a coefficient $(k^2 - k_1^2)/k$ and a summand proportional to the surface impedance Z. It is impossible to reduce this expression to tabulated functions in calculating the mutual impedance of two identical radiators, since the integrand depends on two propagation constants – k and k_1. This can be done only for calculating the intrinsic impedance of a thin radiator, whose radius is much smaller than the antenna length L and the wavelength λ.

If to introduce a quantity δ, satisfying the inequality $a << \delta << L, \lambda$, then $R = \varsigma - z$, when $|\varsigma - z| > \delta$, and $\exp(-jkR) = 1$, when $|\varsigma - z| > \delta$. By dividing the integration interval into intervals from $-L$ to $-\delta$, from $-\delta$ to 0, from 0 to δ and from δ to L, and sequentially using the obtained approximate magnitudes and (1.8), we arrive at an approximate expression for the input impedance of an impedance radiator:

$$R_A = \frac{30}{\sin^2\beta}\left\{\left(\frac{\beta}{\alpha} + \frac{\alpha}{\beta}\right)\left[\ln\frac{m}{l} - Cim + Cil + \frac{1}{2}\cos 2\beta\left(\ln\frac{m}{l} - 2Cim + 2Cil + Ci2m - Ci2l\right) + \frac{1}{2}\sin 2\beta(Si2m - Si2l - 2Sim + 2Sil)\right] -$$

$$\frac{\beta^2 - \alpha^2}{\alpha}[Sim - Sil + \cos 2\beta(Sim - Sil - Si2m + Si2l) + \sin 2\beta(Ci2m - Ci2l - Cim + Cil)] - 2(\cos\beta - \cos\alpha)^2\},$$

$$X_A = \frac{60\beta}{\chi\alpha}\cot\beta + \frac{30}{\sin^2\beta}\left\{\left(\frac{\beta}{\alpha} + \frac{\alpha}{\beta}\right)\left[Sim + Sil + \frac{1}{2}\cos 2\beta(2Sim + 2Sil - Si2m - Si2l) + \frac{1}{2}\sin 2\beta\left(Ci2m + Ci2l - 2Cim - 2Cil + \ln\frac{\gamma^2 ml}{4\alpha^2}\right)\right] -$$

$$\frac{\beta^2 - \alpha^2}{\alpha}\left[Cim + Cil + \cos 2\beta(Cim + Cil - Ci2m - Ci2l) + \sin 2\beta(-Si2m - Si2l + Sim + Sil) - \ln\frac{\gamma^2 ml}{4\alpha^2}\right] - 2\sin\alpha(2\cos\beta - \cos\alpha) + \frac{\beta}{2}\sin 2\beta\right\}.$$

$$\tag{1.21}$$

In these expressions $\alpha = kL, \beta = k_1 L, m = \beta + \alpha, l = \beta - \alpha, Six = \int\limits_x^x\frac{\sin u}{u}du$ is integral sine, $Cix = \int\limits_\infty^x\frac{\cos u}{u}du$ is integral cosine, $\ln\gamma = C = 0.5772\ldots$ is Euler's constant. Expressions (1.21) were used for calculating input impedances shown in Fig. 1.2.

Integral formulas obtained with a help of Leontovich's integral equation and the induced emf method in the second approximation are identical.

Consequently, the results found in both cases by means of numerical integration are same. Similar results are obtained when using calculation programs based on the moments method.

At a small radiator length, the radiation resistances of the impedance and metal antennas are same.

1.3 Radiators with surface impedance changing along the antenna

An antenna with constant surface impedance is a particular case of a radiator with impedance changing along the antenna. From general considerations, it is clear that if the constant impedance allows expansion of antenna possibilities, then its changing along the antenna will allow greater achievement. Such a radiator can be realized as a structure from $2N$ sections with length l_i and different magnitudes of the surface impedance Z_i at the sections. But on each section, a surface impedance is constant (Fig. 1.4). We will assume that the antenna radius a is much smaller than the wavelength

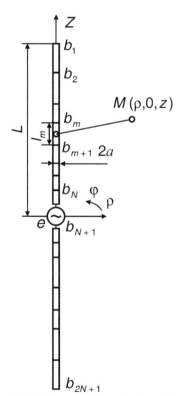

Fig. 1.4: Radiator with changing impedance.

λ and the length of the antenna $2L$. Let us show that in this case the current distribution along each antenna section has a sinusoidal character in the first approximation.

The vector potential of a harmonic field is equal to

$$A_z(\rho,z) = \frac{\mu}{8\pi^2} \sum_{i=1}^{2N} \int_0^{2\pi} \int_{b_{i+1}}^{b_i} \frac{J_i(\varsigma)e^{-jkR}}{R} d\varphi d\varsigma, \tag{1.22}$$

where $R = \sqrt{(z-\varsigma)^2 + \rho^2 + a^2 - 2a\rho\cos\varphi}$ is the distance from an observation point M with coordinates $(\rho, 0, z)$ to an integration point P with coordinates (a, φ, ς). Calculating an internal integral of expression (1.22) by parts and taking into account the current continuity along the antenna and its absence at its ends:

$$J_i(b_i) = J_{i-1}(b_i), J_1(b_1) = J_{2N}(b_{2N+1}) = 0, \tag{1.23}$$

we get

$$A_z(\rho,z) = \frac{\mu}{4\pi}\left[-2J_n(z)\ln p\rho + \sum_{i=1}^{2N} V(J_i, z)\right], b_{n+1} \le z \le b_n, \tag{1.24}$$

where $V(J_i, z) = \int_{b_{i+1}}^{b_i} e^{-jk|z-\varsigma|} \ln 2p(z-\varsigma)\left[jkJ_i(\varsigma) + sign(z-\varsigma)\frac{dJ_i(\varsigma)}{d\varsigma}\right]d\varsigma.$

Substituting (1.24) into the expression (1.19) for the field $E_z(\rho, z)$ in a cylindrical coordinate system and equating $\rho = a$, we find the tangential component $E_z(a, z)$ on the antenna surface. In accordance with the boundary condition (1.1) for the m-th section, we obtain the equation

$$\frac{d^2 J_m(z)}{dz^2} + k^2 J_m(z) = -4\pi j\omega\varepsilon_0\chi\left[K(Z) + \sum_{i=1}^{2N} W(J_i) - \frac{J(z)Z_m}{2\pi a}\right], b_{m+1} \le z \le b_m, \tag{1.25}$$

where $4\pi j\omega\varepsilon_0 W(J_i) = \frac{d^2 V(J_i, z)}{dz^2} + k^2 V(J_i, z)$, and Z_m is a value of the surface impedance on the m-th section.

As in Section 1.1, we will consider such values of the impedances, which significantly affect the current distribution. Then

$$\frac{d^2 J_m(z)}{dz^2} + k_m^2 J_m(z) = -4\pi j\omega\varepsilon_0\chi\left[K(Z) + \sum_{i=1}^{2N} W(J_i)\right], \tag{1.26}$$

where $k_m = \sqrt{k^2 - j2k\chi Z_m/(aZ_0)}$ is the wave propagation constant along the m-th section of the antenna. We seek the solution of (1.26) in the form of a series in powers of a small parameter, which is similar to a series (1.5):

$$J_m(z) = \sum_{n=0}^{\infty} \chi^n J_{mn}(z),$$
(1.27)

that allows us to reduce (1.26) to the system of equations

$$\frac{d^2 J_{m1}(z)}{dz^2} + k_m^2 J_{m1}(z) = -4\pi j\omega\varepsilon_0 K(z),$$

$$\frac{d^2 J_{mn}(z)}{dz^2} + k_m^2 J_{mn}(z) = -4\pi j\omega\varepsilon_0 \sum_{i=1}^{2N} W\left(J_{i,n-1}\right), n > 1, \ b_{m+1} \le z \le b_m, \quad (1.28)$$

and the boundary conditions for $J_{mn}(z)$ are analogous to the boundary conditions for $J_m(z)$. If the values of the current $J_{m1}(b_m)$ and $J_{m1}(b_{m+1})$ at the ends of m-th section are given and the radiator is excited by the concentrated emf, included at a point $z = h$, we find from the first equation

$$\chi J_{m1}(z) = \chi J_{m1}(b_m) \frac{\sin k_m(z - b_{m+1})}{\sin k_m l_m} + \chi J_{m1}(b_{m+1}) \frac{\sin k_m(b_m - z)}{\sin k_m l_m} +$$

$$j \frac{\Gamma_1 k \chi e}{30 k_m \sin k_m l_m} [\Gamma_2 \sin k_m(h - b_{m+1}) \sin k_m(b_m - z) + (1 - \Gamma_2) \sin k_m(z - b_m + 1)],$$
(1.29)

where $\Gamma_1 = \begin{cases} 1, b_{m+1} \le h \le b_m, \\ 0, h > b_m, h < b_{m+1}, \end{cases} \quad \Gamma_2 = \begin{cases} 1, h \le z, \\ 0, h > z. \end{cases}$

As can be seen from (1.29), the current in the first approximation is equal to the sum of the first two terms in all the sections, except for the section where the concentrated emf is placed. If, for simplicity, the emfs are placed only at the boundaries of the sections, then, as can be easily verified, on the sections adjacent to the generator, the additional term is equal to zero, i.e.,

$$\chi J_{m1}(z) = I_m \sin(k_m z_m + \varphi_m),$$
(1.30)

where $I_m = \dfrac{\chi}{\sin k_m l_m} \sqrt{J_{m1}^2(b_m) + J_{m1}^2(b_{m+1}) - 2J_{m1}(b_m) J_{m1}(b_{m+1}) \cos k_m l_m},$

$$\varphi_m = \tan^{-1} \frac{J_{m1}(b_m) \sin k_m l_m}{J_{m1}(b_{m+1}) - J_{m1}(b_m) \cos k_m l_m}, z_m = b_m - z.$$

Thus, in the antenna with a stepped change of the surface impedance, the current distribution in each section has a sinusoidal character. In order to find the law of current distribution along the entire radiator, it is necessary to add to a law of current continuity (1.23) a law of charge continuity. In accordance with the equation of charge continuity at the sections' boundaries

$$divj = -\partial\rho/\partial t, \tag{1.31}$$

where j is a density of a conduction current, and ρ is a density of an electric charge, we get

$$\frac{dJ_m(b_m)}{dz} = \frac{dJ_{m-1}(b_m)}{dz}. \tag{1.32}$$

This equality holds for all m, except the point $z = h = b_N$ of a generator's location, where difference of the derivatives on the right and on the left is equal to

$$\frac{dJ_N(h+0)}{dz} - \frac{dJ_N(h-0)}{dz} = 2\frac{dJ_N(h+0)}{dz} = -j\frac{k\chi e}{30k_N}.$$

Assuming that the radiator is symmetric, we obtain from (1.23), (1.30) and (1.32)

$$I_m \sin\varphi_m = I_{m-1}\sin(k_{m-1}l_{m-1} + \varphi_{m-1}), \, I_m k_m \cos\varphi_m = I_{m-1}k_{m-1}\cos(k_{m-1}l_{m-1} + \varphi_{m-1}).$$

These equalities allow us to express the amplitude and phase of the current in any section through the amplitude and phase of the current of the previous section

$$I_m = I_{m-1}\frac{\sin(k_{m-1}l_{m-1} + \varphi_{m-1})}{\sin\varphi_m}, \tan\varphi_m = \frac{k_m}{k_{m-1}}\tan(k_{m-1}l_{m-1} + \varphi_{m-1}), \quad (1.33)$$

and consequently, through the parameters of the sections and one of the currents:

$$I_m = I_N \prod_{p=m+1}^{N} \frac{\sin\varphi_p}{\sin(k_{p-1}l_{p-1} + \varphi_{p-1})},$$

$$\varphi_m = \tan^{-1}\left\{\frac{k_m}{k_{m-1}}\tan\left[k_{m-1}l_{m-1} + \tan^{-1}\left\{\frac{k_{m-1}}{k_{m-2}}\tan\left(k_{m-2}l_{m-2} + ... + \tan^{-1}\left(\frac{k_2}{k_1}\right)...\right)\right\}\right]\right\}.$$

$$\tag{1.34}$$

The last expression will be also true for the N-th section, if to come to an agreement that the product $\prod\limits_{p=N+1}^{N}$ is equal to 1.

Since the current of the generator is $J(0) = J_N \sin(k_N l_N + \varphi_N)$, then

$$I_m = A_m J(0), \tag{1.35}$$

where $A_m = \prod_{p=m+1}^{N} \sin \varphi_p \Big/ \prod_{p=m}^{N} \sin(k_p l_p + \varphi_p)$.

Expression (1.30), together with (1.34), is the required law of a current distribution along the radiator. This distribution can also be obtained in another simpler way, if the concept of an impedance long line, open at the end is used. For a changing impedance (more precisely, piecewise-constant impedance), this line is non-uniform; more precisely, it is a stepped long line. It consists of N uniform sections of length l_m with wave impedance W_m, current J_m and voltage u_m (Fig. 1.5).

Comparison of the sections of the impedance radiator and the stepped line allows occurrence of expressions for the propagation constant k_m and the wave impedance W_m analogous to (1.7) and (1.10). On the basis of the theory of long lines, it is known that the voltage and current along a section of a uniform line are equal to

$$u_m = U_m \cos (k_m z_m + \varphi_m), \quad J_m = jI_m \sin (k_m z_m + \varphi_m), \tag{1.36}$$

and $I_m = U_m/W_m$. Since the voltage and current along the equivalent stepped line are continuous:

$$u_m\big|_{z_m=0} = u_{m-1}\big|_{z_{m-1}=l_{m-1}}, J_m\big|_{z_m=0} = J_{m-1}\big|_{z_{m-1}=l_{m-1}},$$

then

$$I_m = I_{m-1} \frac{\sin(k_{m-1}l_{m-1} + \varphi_{m-1})}{\sin \varphi_m}, U_m = U_{m-1} \frac{Cos(k_{m-1}l_{m-1} + \varphi_{m-1})}{\cos \varphi_m}. \tag{1.37}$$

Dividing the first of these expressions into the second one, we find, taking into account (1.36):

$$\tan \varphi_m = \frac{k_m}{k_{m-1}} \tan(k_{m-1}l_{m-1} + \varphi_{m-1}). \tag{1.38}$$

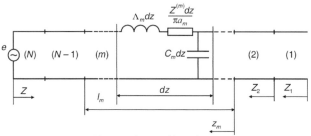

Fig. 1.5: Stepped long line.

The equalities (1.37) and (1.38) are identical to the equalities (1.33), i.e., along the stepped long line the same law of current distribution is valid. Since the generator emf and current are equal, respectively, to $e = U_N \cos(k_N l_N + \varphi_N)$ and $J(0) = jI_N \sin(k_N l_N + \varphi_N)$, the input impedance of the stepped line is

$$Z_l = e/J(0) = -jW_N \cot(k_N l_N + \varphi_N). \tag{1.39}$$

The input impedance of the impedance radiator is equal in the first approximation to the input impedance Z_l of the impedance line, i.e., it is purely reactive similarly to the purely reactive input impedances of the metallic radiator and the antenna with a constant surface impedance. To calculate the magnitude of a resistance and to define more exactly the reactance magnitude, it is necessary to solve the problem in the second approximation. In this case, it is necessary to apply numerical integration, i.e., use programs based on the moment method, or generalize the expression (1.20) to the case of changing surface impedance. This generalized expression has the form:

$$Z_A = j60 \sum_{m=1}^{N} A_m \int_{b_{m+1}}^{b_m} \left\{ \sum_{i=1}^{N} A_i \left(k - \frac{k_i^2}{k} \right) \int_{b_{i+1}}^{b_i} \left(\frac{e^{-jkR}}{R} + \frac{e^{-jkR_+}}{R_+} \right) \sin[k_i(b_i - \varsigma) + \varphi_i] d\varsigma + \right.$$

$$\frac{A_1 k_1}{k} \left(\frac{e^{-jkR_1}}{R_1} + \frac{e^{-jkR_2}}{R_2} \right) - \frac{2A_N k_N}{k} \cos(k_N l_N + \varphi_N) \frac{e^{-jkR_0}}{R_0} \left. \right\} \sin[k_m(b_m - z) + \varphi_m] dz + \tag{1.40}$$

$$\frac{1}{2\pi a} \sum_{m=1}^{N} A_m^2 Z_m \left[l_m - \frac{1}{k_m} \sin k_m l_m \cos(k_m l_m + 2\varphi_m) \right].$$

In Fig. 1.6 the reactive component of input impedance and the radiation resistance of two asymmetrical antennas with a ferrite shell and following geometric dimensions (in meters) $L = 2.0, a_1 = 0.007, a = 0.021$ are presented. The relative ferrite permittivity is close to 10 and the relative magnetic permeability as a function of z coordinate is shown in this figure. In Fig. 1.6a the shell covers the upper part of the metal rod and in Fig. 1.6b, the lower part. The calculations are performed in accordance with (1.40). As can be seen from the figures, in the second variant, the series resonance is shifted towards low frequencies. The experimental values are given by circles and triangles. The coincidence of the experiment with the calculation is satisfactory.

Let us give also an expression for the impedance radiator field in the far zone. This field is not difficult to determine, if the radiator is considered as a sum of elementary electric dipoles located along z axis with the antenna center at the coordinates origin:

$$E_\theta = j30k \frac{\exp(-jkR)}{R} \sin \theta \int_{-L}^{L} J(z) \exp(jkz \cos \theta) dz, \tag{1.41}$$

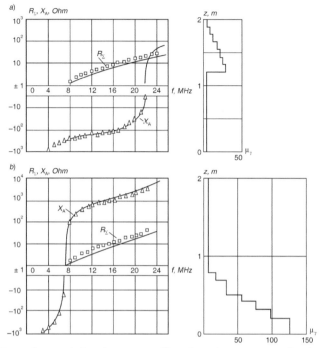

Fig. 1.6: Input characteristics of antennas with a piecewise constant surface impedance.

where R is the distance from the center of the antenna to the point of observation, $kz\cos\theta$ is the path difference between the points z and 0 of the antenna. If the current distribution along the antenna is determined by the expression (1.35), then

$$E_\theta = j30kJ(0)\frac{\exp(-jkR)}{R}F(\theta), \tag{1.42}$$

where

$$F(\theta) = \sin\theta\sum_{m=1}^{N} A_i \left[\frac{\cos(\varphi_m - kb_m\cos\theta) - \cos(\varphi_m + k_m l_m - kb_{m+1}\cos\theta)}{k_m + k\cos\theta} + \right.$$

$$\left. \frac{\cos(\varphi_m + kb_m\cos\theta) - \cos(\varphi_m + k_m l_m - kb_{m+1}\cos\theta)}{k_m - k\cos\theta} \right], \tag{1.43}$$

If the surface impedance is constant along the antenna, then its field in the far zone is equal to

$$E_\theta = j60J(0)\frac{\exp(-jkR)}{R}\frac{k_1 k\sin\theta[\cos(kL\cos\theta) - \cos k_1 L]}{(k_1^2 - k^2\cos^2\theta)\sin k_1 L}. \tag{1.44}$$

1.4 How do mistakes arise?

The theoretical and experimental results described in the previous sections were obtained mainly in the sixties of the last century. A stimulus of these works became the article [1] about an antenna in the form of a metal rod with a ferrite shell. This article clearly showed that the ferrite coating changes the propagation constant of a wave along the antenna and the current distribution along its axis. This fact changes all electrical characteristics of the antenna. The obtained results allowed creation of a theory of antennas with both constant and changing surface impedance [4-6]. But these results did not lead to a serious expansion of the operating frequency band of antennas, since the question of the rational change in the impedance along the antenna for decreasing input reactance and improvement of an antenna matching with a cable in a wide frequency band remained open. The reason of this circumstance was incorrect selection of surface impedance and equivalent loads. This history deserves attention.

The calculations showed that a change of the surface impedance at any section changes the resonant frequencies of an antenna. In order to determine the effect of antenna elements on its input characteristics, an antenna with a constant frequency of the first (series) resonance was considered and its input impedance was determined for different variants of changing surface impedance along an antenna [6].

Input reactance of such an antenna in accordance with (1.34) is equal to

$$X_A = 60k_0 X/(k\chi),\tag{1.45}$$

where $X = -\dfrac{k_N}{k_0}\cot\left\{Mk_N l_N + \tan^{-1}\left[\dfrac{k_N}{k_{N-1}}\tan\left(Mk_{N-1}l_{N-1}+\ldots+\tan^{-1}\left(\dfrac{k_2}{k_1}\right)\right)\right]\right\}$

is the reactance of an antenna, reduced to a certain magnitude, which depends only on the ratio of different propagation constants, but not on their absolute magnitude. In this expression k_m is the propagation constant in the arbitrary m-th section with length l_m, k_0 is the propagation constant for the antenna with a constant surface impedance and is equal to $k_0 = \pi/(2L)$, M is the relative frequency and is equal to $M = f/f_1$, f is the frequency, f_1 is the frequency of the first resonance. The expression for X contains magnitude $Mk_m l_m = \dfrac{k_m}{k_0}\cdot\dfrac{Ml_m\pi}{2L}$. This magnitude is equal to the product of a ratio k_m/k_0 and a constant factor of the m-th section. Thus, at a given relative frequency M the reactance X depends only on the ratio of the magnitudes k_i/k_0. These magnitudes are related to each other by the condition of invariance in the frequency of the first resonance:

$$X_A\big|_{M=1} = X\big|_{M=1} = 0.\tag{1.46}$$

Since the resulting equation is transcendental, we will consider it successively for radiators with a different number of sections of the same length. Let's start with two sections: first, in accordance with (1.46), we calculate the magnitude k_2/k_0 as a function of k_1/k_0. This dependence is given in Fig. 1.7. In Fig. 1.8 the curves for the reactance X are plotted depending on frequency M for each point of Fig. 1.7. From the Fig. 1.8 one can be seen that at $M > 1$ curves pass closer to the abscissa axis, if $k_1 < k_0 < k_2$. The best option is when $k_1 = 0$. This means that on the section near the free end of the antenna, the surface impedance should be zero.

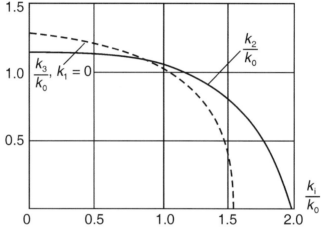

Fig. 1.7: The relationship between propagation constants of different sections at $X_A|_{M=1} = 0$.

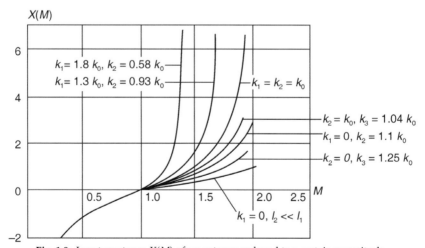

Fig. 1.8: Input reactance $X(M)$ of an antenna reduced to a certain magnitude.

Proceeding on the basis of this result, in an antenna of three sections of equal length, we can put $k_1 = 0$. For this variant in Fig. 1.7, the ratio k_3/k_0 is presented as a function of k_2/k_0, and in Fig. 1.8, the curves are given for the reactance $X(M)$. The calculation again shows that one must concentrate the surface impedance near the generator. Therefore, as a next step, it is expedient to consider the radiator from two sections of different lengths l_1 and l_2, where $l_1 \approx L$, $l_2 \ll l_1$, $k_1 \ll k_0$. Analysis again shows that the lower reactance $X(M)$ is obtained for the smaller length l_2. This result corresponds to a conventional wire antenna with a loading coil at the base.

Rigorous analysis reveals the obvious incorrectness of an obtained result, but the main reason for the error becomes evident only with time. In the given case, when calculating reactive impedances, it was assumed that the surface impedance is positive, i.e., it has an inductive character and the currents are distributed according to the sinusoidal law.

Manufacture of wide-range radiators providing a high level of matching with the cable and a maximum radiation in a plane perpendicular to the radiator axis is based on the creation of an in-phase current in the radiator. The current amplitude varies along the radiator axis in accordance with a linear or exponential law. In order to obtain such current distributions, it is necessary to include, along the radiator axis, concentrated capacitive loads, the magnitude of which varies along this axis. As it follows from the above, the method of solution proposed in [6] had little in common with the solution that led to a correct result.

This is not the only case when the usual ideas prevent us from explaining the not too-complicated contradictions. Using the first formulation of the emf method for calculation of the loss in resistance of antennas (of a loss to skin effect, of a loss in ferrite) led to an unexpected result, viz. to negative resistance. The accepted second formulation has three advantages: the coincidence of analytic expressions with the results of solving integral equations, the positive resistance of losses and the stability of a solution. Stability of the solution means that changing the initial parameter by a value of the first order of smallness will lead to a change in the result by a value of the second order of smallness.

The validity of this proposition was quickly proved for the second formulation of the method of induced emf, but it was not possible to prove this statement with respect to the first formulation. In trying to prove the unprovable proposition, i.e., the stability of the first formulation, the well-known scientist published the required proof. Unfortunately, he made mistakes in the first line of his proof, assuming that the current in the general case is an arbitrary but purely reactive quantity. Such an assumption was unacceptable, if only because the obviously erroneous results took place when analyzing a structure with complex characteristics and the losses created an active component of a current.

One more example is the analysis of the influence of loads on the shape of an antenna's directional pattern. Considering the problem of synthesizing antennas with loads, my opponent studied the directional patterns of antenna, consistently increasing the number of loads in the radiator. After switching on the first load, the directional pattern changed significantly; after switching on the second load it changed slightly and after switching on the third load, the change became invisible. He published an article in which he concluded that the load does not affect directional pattern. In this work two mistakes were made: first, the comparison of pictures without their mathematical processing is unacceptable; secondly, if to increase the number of loads to ten, the quantity will pass into quality and the effect of loads on the radiation pattern will be very appreciable.

Paradoxes of this type, which sometimes lead to errors of well-known scientists, are discussed in detail in the book [7]. They require a close study of the results and a careful approach to conclusions.

Radiators with Concentrated Loads

2.1 Capacitive loads and distribution of in-phase current along the radiator

Chapter 1 is devoted to antennas with distributed surface impedance. In Chapter 2, antennas with concentrated impedance, or in other words, with concentrated loads, are considered. An example of such an antenna with concentrated loads can serve an antenna with capacitors as shown in Fig. 2.1. An integral equation for a current in such an antenna is easily obtained from an integral equation for a current in a metal radiator. The dimensions of the capacitors for simplicity are assumed to be infinitesimal. The inclusion of a capacitor is a special case of inclusion of a concentrated complex impedance Z_n at the point $z = z_n$, which is equivalent to an inclusion at this point of an additional concentrated emf $e_n = -J(z_n)Z_n$, where $J(z_n)$ is the current at the point z_n. The external field, which corresponds to this emf, is equal to

$$E_n = -J(z_n)Z_n\delta(z - z_n). \qquad (2.1)$$

In the case of N loads, the boundary condition for the tangential component of the electric field on the radiator surface has the form

$$E_z(J)\big|_{\rho=a,-L\leq z\leq L} + K(z) - \sum_{n=1}^{N} J(z_n)Z_n\delta(z - z_n) = 0. \qquad (2.2)$$

Accordingly, instead of equation (1.4), we obtain

$$\frac{d^2J(z)}{dz^2} + k^2 J(z) = -4\pi j\omega\varepsilon_0\chi\left[K(z) + W(J,z) - \sum_{n=1}^{N} J(z_n)Z_n\delta(z - z_n)\right], J(\pm L) = 0 \qquad (2.3)$$

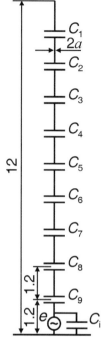

Fig. 2.1: Antenna with capacitors.

If the radiator is symmetrical and the loads of its arms are same and placed at equal distances from the antenna center, then from (2.2) it follows:

$$\frac{d^2 J(z)}{dz^2} + k^2 J(z) = -4\pi j\omega\varepsilon_0 \chi \left[K(z) + W(J,z) - \sum_{n=1}^{N} J(z_n) Z_n \delta(z - z_n) \right], J(\pm L) = 0 \tag{2.4}$$

For example, Gallen's equation for the filament takes the form:

$$\int_{-L}^{L} J(\varsigma) G_2 d\varsigma = -j\frac{1}{Z_0} \left\{ C \cos kz + \frac{e}{2}\sin k|z| - \frac{1}{2}\sum_{n=1}^{N/2} J(z_n) Z_n \left[\sin k|z - z_n| + \sin k|z + z_n| \right] \right\}. \tag{2.5}$$

This equation was used in [8]. Here G_2 is the Green's function equal to $G_2 = \frac{\exp(-jkR)}{4\pi R}$, $R = |z - \varsigma|$. The function $J(z_n)$ must satisfy the condition of zero current at the ends of the antenna

$$J(\pm L) = 0. \tag{2.6}$$

As can be seen from (2.3) and (2.5), if the concentrated impedance is included in the antenna, then in the integral equation together with the integral, a free term appears. It is proportional to the magnitude of this impedance. An analogous free term appears in the equation solution. It is also proportional to this impedance and to the square of the current at the inclusion point. It should be emphasized that the inclusion of loads leads not only to the appearance of a free term, but also to a change of integral terms depending on the current distribution. Both, a free term and a change of integral terms, essentially vary the solution of each equation, i.e., current distribution, input and directional characteristics. This means that similarly to the external fields (exciters), the loads can be used to change current distribution and electrical characteristics of an antenna in the desired direction.

Unfortunately, in similar cases, the incorrect logic of research is widely used. Such logic is described at the end of the previous chapter on the example of analysis affecting concentrated loads to directional characteristics of antennas. This logic is fundamentally incorrect since, with a multiple increase of loads number, the quantity can turn into quality.

If the loads number is large enough and they are located at small electrical distances from each other, one can consider that the loads are included along the entire length of the antenna. This means conversion of a radiator with a finite number of loads into an impedance radiator. This radiator, in contrast to the metal one (without loads), has additional degrees of freedom.

As is shown in the book [9], with the help of loads it is possible to solve the inverse problem of a radiators theory—to obtain an antenna with the desired electrical characteristics. A special case of this problem—the creation of a radiator that provides a high level of matching in a wide frequency range and a maximum of radiation in a plane perpendicular to the radiator axis—has great practical importance.

A simple linear radiator is thin and does not contain concentrated loads. It does not satisfy these requirements. A reactive component of its input impedance is large everywhere except in the vicinity of the first series resonance. That leads to a weak matching of antenna and cable impedances. When a monopole length increases, radiation in a plane perpendicular to the radiator axis is reduced, since a current distribution along a thin linear radiator without loads (Fig. 2.2a) is close to sinusoidal one and if a length of radiator arm is more than 0.7λ (λ is a wavelength) anti-phase currents are formed along the antenna section (Fig. 2.2b, curve 1).

It is possible to obtain a current distribution which is different from sinusoidal distribution, if to include the concentrated loads along the radiator length [10–14]. Its shape depends on load magnitudes and points of inclusion. The experimental results [11] show that the radiator with an in-phase current distributed along its axis in accordance with linear or exponential law has over a wide range of frequencies good characteristics (a

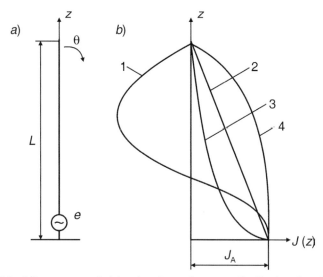

Fig. 2.2: A linear monopole (a) and variants of currents distribution along it (b).

high level of an impedance matching and a required shape of the directional pattern). Such a distribution is created particularly by means of loads with capacitances decreasing in the direction of the free ends of antenna. These results confirm that the maximum radiation is provided in the direction perpendicular to the radiator axis, if the current is in phase over the entire antenna length. In addition, the long antenna with in-phase current has a large resistance of radiation that allows increasing matching level.

Let N concentrated load Z_n be uniformly placed along an asymmetric radiator of height L at a distance b from each other (Fig. 2.3a). If this distance is small ($kb \ll 1$), then the current distribution along the radiator will practically not change if the concentrated loads are replaced by the surface impedance distributed along the length of each section. We assume that the surface impedance of each section is constant and equal to $Z^{(n)}$. Then

$$Z_n = \frac{b}{2\pi a_n} Z^{(n)}. \qquad (2.7)$$

where a_n is the radius of n-*th* section.

As shown earlier, the current distribution along the antenna with a piecewise constant surface impedance is similar in first approximation to a current distribution along the long line with a stepped change of the propagation constant (Fig. 2.3b). In accordance with (1.7) and (2.7) the wave propagation constant γ_n on the n-*th* section of the line is related to the surface impedance $Z^{(n)}$ and the load Z_n by means of equality

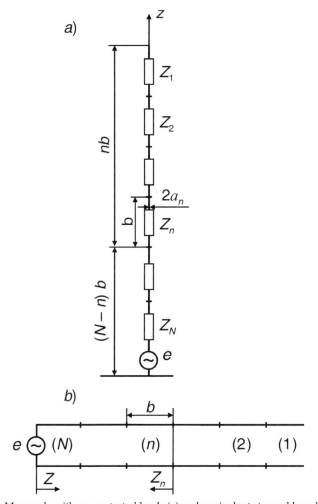

Fig. 2.3: Monopole with concentrated loads (a) and equivalent stepped long line (b).

$$-\gamma_n^2 = k^2 - j\frac{2k\chi_n Z^{(n)}}{a_n Z_0} = k^2 - j\frac{k\chi_n Z_n}{30b},$$
(2.8)

where $\chi = 1/[2\ln(2L/a_n)]$ is the magnitude of a small parameter on the *n-th* section. In a general case, γ_n is a complex value; in a particular case, when this value is purely imaginary ($\gamma_n = jk_n$), the current distribution on the *n-th* section of the line has a sinusoidal character. If we find the law of changing propagation constant γ_n along the radiator that provides the required current distribution, then it is possible by means of (2.7) and (2.8) to calculate the necessary concentrated loads. Indeed, the current on the n-*th* section of the stepped line for an arbitrary γ_n is equal to

$$J(z_n) = I_n sh(\gamma_n z_n + \varphi_n), 0 \le z_n \le b, \tag{2.9}$$

where I_n and φ_n are the amplitude and phase of a current on n-*th* section, and z_n is the coordinate measured from n-th section end, i.e., $z_n = (N - n + 1)b - z$. Suppose the current distribution along the line is equal to

$$J(z) = J_A f(z), 0 \le z \le L, \tag{2.10}$$

where J_A is the current at the line input (the generator current) and $f(z)$ is the real and positive distribution function, corresponding to in-phase current. If, to equate the current $J(z)$ to currents $J(z_n)$ at the ends of each section, then for small distances b the current distribution along the line is close to the required one.

In accordance with (2.9) and (2.10), for $z_n = b$ $I_n sh(\gamma_n b + \varphi_n) = J_A f[(N - n)b]$ and for $z_n = 0$ $I_n sh\varphi_n = J_A f[(N - n + 1)b]$. Divide the left and right parts of the first equality on to the respective parts of the second one and confine by the first terms of expansion of hyperbolic functions with small arguments into series (considering that b is a small magnitude). Then

$$\tanh \varphi_n = \gamma_n b \bigg/ \left\{ \frac{f[(N - n)b]}{f[(N - n + 1)b]} - 1 \right\}. \tag{2.11}$$

For (n+1)-*th* section, similarly to (2.11)

$$\tanh \varphi_{n+1} = \gamma_{n+1} b \bigg/ \left\{ \frac{f[(N - n - 1)b]}{f[(N - n)b]} - 1 \right\}. \tag{2.12}$$

Since a voltage and a current are continuous along the stepped line, then

$$\tanh \varphi_{n+1} = (\gamma_{n+1}/\gamma_n)\tanh(\gamma_n b + \varphi_n). \tag{2.13}$$

Equations (2.11)–(2.13) present a system of equations that allow connection of γ_n and γ_{n+1}. The solution of this system shows that these magnitudes do not depend on each other:

$$\gamma_n = \frac{1}{b} \sqrt{1 - \frac{2 f[(N - n)b] - f[(N - n - 1)b]}{f[(N - n + 1)b]}}. \tag{2.14}$$

At linear distribution of the in-phase current amplitude (Fig. 2.2b, curve 2)

$$J_2(z) = J_A(1 - z/L), \tag{2.15}$$

where

$$f_2(z) = (L-z)/L = \left[(n-1)b + z_n\right]/(Nb).$$

The linear distribution is a particular case of the exponential one (Fig. 2.2b, curves 3 and 4):

$$J_{3,4}(z) = J_A \frac{\exp(-\alpha z) - \exp(-\alpha L)}{1 - \exp(-\alpha L)},$$

that is

$$f_{3,4}(z) = \frac{\exp(-\alpha z) - \exp(-\alpha L)}{1 - \exp(-\alpha L)} = \frac{\sinh\{(\alpha/2)[(n-1)b + z_n]\}}{\sinh(\alpha Nb/2)}, \quad (2.16)$$

where α is logarithmic decrement. If α is positive, a current curve is concave, i.e., the current quickly decreases from the maximum value near the generator to zero. If α is negative, the curve of the current is convex, i.e., the current is more evenly distributed along the radiator. The steepness of the curve depends on the value of α. It is easy to verify that if α tends to zero, $J_{3,4}(z)$ turns into $J_2(z)$.

Note the two special cases of an exponential distribution: when $\alpha \to -\infty$, the current distribution along the antenna becomes rectangular; when $\alpha \to \infty$, the current rapidly decreases with increasing z and is equal to zero everywhere, except for the area near the generator.

The use of exponential distribution allows us to cover a wide class of distributions. Substituting (2.16) into (2.14), we obtain

$$\gamma_n = \frac{\alpha}{\sqrt{2}}\sqrt{1 + \coth[\alpha(n-1)b/2]}. \quad (2.17)$$

In particular, for $0 < \alpha << 1/L$

$$\gamma_n = \sqrt{\alpha/[(n-1)b]}. \quad (2.18)$$

Thus, in order to obtain a concave exponential current distribution along the stepped line, the propagation constant must be real and vary along the line axis according to (2.17). In particular, in order to obtain a distribution close to linear, the quantity γ_n must be small and inversely proportional to square root of a distance from a free end of the line. For $\alpha < 0$

$$\gamma_n = j\frac{|\alpha|}{\sqrt{2}}\sqrt{\coth[|\alpha|(n-1)b/2] - 1}, \quad (2.19)$$

i.e., in order to obtain a convex exponential distribution, the propagation constant must be purely imaginary along the entire line. For large positive α,

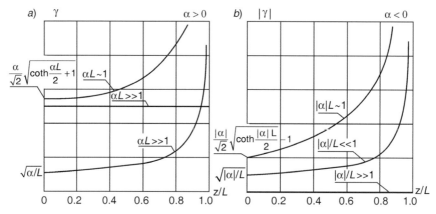

Fig. 2.4: The propagation constant for positive (a) and negative (b) logarithmic decrements.

the quantity γ_n is real, is constant along the line length and is equal to a decrement. For large negative α the propagation constant tends to zero. The family of curves for γ as a function of z is given in Fig. 2.4.

Directional pattern of an antenna with in-phase current in a vertical plane in accordance with (1.41) has the form

$$F(\theta) = \sin\theta \int_{-L}^{L} f(z)\exp(jkz\cos\theta)dz. \qquad (2.20)$$

As is clear from (2.10), function $f(z)$ characterizes the law of changing current amplitude along the radiator. For an exponential distribution, the calculation of integral leads to the expression

$$F_{3,4}(\theta) = \frac{\tan\theta}{k^2\cos^2\theta + \alpha^2}\{k\cos\theta - e^{-\alpha L}[k\cos\theta\cos(kL\cos\theta) + \alpha\sin(kL\cos\theta)]\}. \qquad (2.21)$$

In particular, for linear distribution

$$F_2(\theta) = (\sin\theta/\cos^2\theta)[1 - \cos(kL\cos\theta)]. \qquad (2.22)$$

Figure 2.5 demonstrates the normalized antenna patterns with a linear and exponential distribution of current amplitudes. The antenna arm length changes from $3\lambda/4$ up to 4λ. For the sake of clarity, the curves are plotted in a rectangular coordinate system. Here, for comparison, the directional patterns $F_1(\theta)$ of a radiator with a sinusoidal current distribution are presented also. As is seen from Fig. 2.5, for a linear and exponential distribution, unlike a sinusoidal one, the radiation maximum with increasing frequency does not deviate from the perpendicular to the radiator axis. The increase of

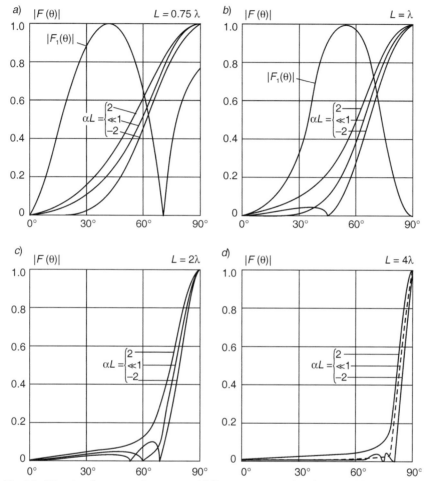

Fig. 2.5: Directional patterns of antennas with linear, exponential and sinusoidal distributions of a current amplitude for different arm lengths: $3\lambda/4$(a), λ (b), 2λ (c), 4λ(d).

L/λ tapers the main lobe and increases the directivity. The width of the main lobe also decreases with decreasing α, including negative values of α.

The antenna input impedance in the first approximation is equal to the input impedance of the stepped long line, i.e.,

$$Z_l = -jW_N \coth{(\gamma_n b + \varphi_n)}|_{n=N}. \qquad (2.23)$$

In this expression, unlike (1.39), the trigonometric cotangent is replaced by a hyperbolic one. The wave impedance of the line is equal to $W_N = (\gamma_n/k)W$, where W is the wave impedance of the metal radiator with the same dimensions, but without loads.

Using the equalities (2.12) and (2.13), we find:

$$\frac{1}{b}\left\{\frac{f[(N-n-1)b]}{f[(N-n)b]}-1\right\}=\frac{\gamma_n}{\tan(\gamma_n b+\varphi_n)},$$

from where

$$Z_l=-j\frac{W}{k}\frac{\gamma_n}{\tan(\gamma_n b+\varphi_n)}\Big|_{n=N}-j\frac{W}{kb}\left\{\frac{f[(N-n-1)b]}{f[(N-n)b]}-1\right\}\Big|_{n=N}.$$

Substituting $n = N$, we get

$$Z_l = -j(W/kb)[f(-b)-1]. \tag{2.24}$$

From this expression, a very important conclusion follows: in order to reduce the reactive component of the radiator input impedance, the function $f(z)$ should slowly change near the antenna base. Then the value in square brackets will be a small value to an order of kb. Otherwise, the reactive component increases sharply.

For an exponential distribution, replacing $f(-b)$ by $f_3(-b)$, we get:

$$Z_{l3} = -j(W/kL)f_x(\alpha L/2), \tag{2.25}$$

where $f_x(x) = x(1 + \coth x)$. The curve of function $f_x(x)$ is given in Fig. 2.6a. In particular for a linear distribution $f_x(x) = 1$, i.e.,

$$Z_{l2} = -jW/(kL). \tag{2.26}$$

In Fig. 2.6b the input reactance X_{A1} of the uniform line with the sinusoidal current distribution and the input reactance X_{A3} of a non-uniform

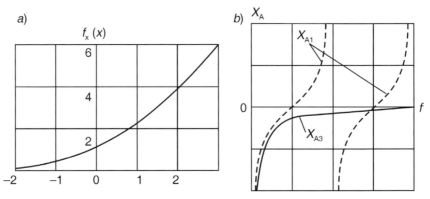

a)

b)

Fig. 2.6: Curve of function $f_x(x)$ (a) and the input impedances of uniform and non-uniform line (b).

line with exponential current distribution are compared with each other, depending on the frequency. The quantity α is assumed to be constant. If in the first case the curve is a cotangent curve, then in the second case it smoothly approaches the axis with increasing frequency, providing high impedance matching over a wide range.

As is seen from (2.25) for Z_{13} and from Fig. 2.6a, the reactance of the line at a given frequency decreases if α decreases. This means that a decrease in α leads to a decrease in the antenna reactance. At the same time, its effective height increases as the area bounded by the curve of a current increases. Consequently, the radiation resistance also increases. Thus, in the case of exponential distribution, it is advisable to decrease α including the negative values.

2.2 Creating in-phase current using the method of impedance line

In the previous section, advantages of in-phase distribution with different decrements α were demonstrated. The positive values of α correspond to the real propagation constant γ_n, and the negative values correspond to a purely imaginary constant γ_n. The possibility of realizing a certain constant (and therefore a certain in-phase distribution) depends on feasibility of the surface impedance or equivalent concentrated loads corresponding to this constant propagation.

To create in-phase current distribution, the magnitude γ_n must be purely real or purely imaginary along an entire antenna. To create in-phase distribution, which will exist over a wide frequency range, the quantity γ_n must be purely real, i.e.,

$$\gamma_n^2 > 0. \tag{2.27}$$

Indeed, as can be seen from (2.9), if the values of γ_n u φ_n are purely imaginary, then the hyperbolic sine becomes trigonometric and with increasing frequency, it will replace the sign when the argument exceeds π. If γ_n is real, then, as is clear from (2.11), for a monotonically decreasing function the sign of φ_n coincides with the sign of γ_n. As it follows from (2.14), for real γ_n

$$f[(N-n)b] \leq 0.5 \{f[N-n+1)b] + f[(N-n-1)b]\}, \tag{2.28}$$

i.e., the function $f(z)$ cannot be convex.

As can be seen from (2.8), the magnitude γ_n^2 consists of two summands. Two possible options for fulfillment of condition (2.27) in wide frequencies range follow from (2.8). In the first case the first summand is small in comparison with the second one:

$$k^2 \ll j \frac{k\chi_n Z_n}{30b}, \tag{2.29}$$

i.e., the square of propagation constant is equal to

$$\gamma_n^2 = j \frac{k\chi_n Z_n}{30b}. \tag{2.30}$$

In accordance with (2.29), the loads in this case must have capacitive character.

The second option is feasible if the second summand $j \frac{k\chi_n Z_n}{30b}$ is proportional to k^2, i.e., if the signs of both summands are same when the frequency changes. In this case the concentrated load Z_n in accordance with (2.8) and ratio $k = \omega/c$, where c is the light speed, is equal to

$$Z_n = -j \frac{30bk}{\chi_n} \left(1 + \frac{\gamma_n^2}{k^2}\right) = -j\omega|\Lambda_n|, \tag{2.31}$$

i.e., Z_n is a negative inductance

$$|\Lambda_n| = \frac{30b}{\chi_n c} \left(1 + \frac{\gamma_n^2}{k^2}\right). \tag{2.32}$$

For a small value of the ratio γ_n/k, the inductance Λ_n depends weakly on the frequency f. In this case the value $\gamma_n^2 = k^2[\chi_n| \Lambda_n| c/(30b)-1]$ is positive, if

$$|\Lambda_n| > 30b/(\chi_n c). \tag{2.33}$$

Negative inductance is a circuit element with negative purely reactive impedance proportional to frequency. This element is equivalent to a frequency-dependent capacitance:

$$-j\omega| \Lambda_n| = 1/(j\omega C_n), \tag{2.34}$$

where

$$C_n = 1/(\omega^2|\Lambda_n|) = C_{n0} f_0^2/f^2.$$

Here C_{n0} is the magnitude of an equivalent capacitance at the frequency $f=f_0$. It does not depend on f. As follows from that, in order to maintain the in-phase distribution of current in a wide frequency range, the capacitances of the concentrated loads must change in inverse proportion to the square of the frequency. If it were possible, the capacitor could be made with the capacitance, which satisfies the condition (2.27) and eliminates the constraint

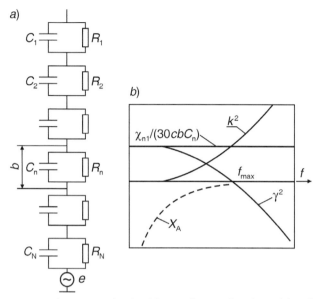

Fig. 2.7: An asymmetric antenna circuit with capacitors and resistors (a) and a frequency dependence of propagation constant (b).

(2.29). But, strictly speaking, purely reactive impedances can only grow, and negative inductances do not exist in Nature. Therefore, the use of constant capacitors allows creating in-phase current distribution only in limited frequency bands.

In the first option, when the condition (2.27) is accomplished, one must refine magnitudes of concentrated loads corresponding to the propagation constant γ_n. If we take into account that the parameter χ_n is in principle a complex quantity ($\chi_n = \chi_{n1} - j\chi_{n2}$), then a load admittance is equal to

$$Y_n = 1/Z_n = j\omega C_n + 1/R_n, \qquad (2.35)$$

and in accordance with (2.30) $C_n = 4\pi\varepsilon_0\chi_{n1}/(b\gamma_n^2)$ and $R_n = b\gamma_n^2/(4\pi\varepsilon_0\omega\chi_{n2})$. Thus, as it follows from (2.35), in order the value γ_n^2 has been real and positive, each load must be executed as a parallel connection of a resistor and a capacitor (Fig. 2.7a). The resistance of the resistor must vary in inverse proportion to the frequency and the capacitance of the capacitor should remain constant. When designing a real antenna, it is advisable to select a value R_n for the average frequency of the range.

To obtain the required current distribution $f(z)$ along the antenna, the magnitude γ_n must correspond to the equality (2.14). Substituting it into (2.35) gives

$$C_n = 4\pi\varepsilon\chi_{n1}b\left\{1 - \frac{2f[(N-n)b] - f[(N-n-1)b]}{f[(N-n+1)b]}\right\}^{-1}, R_n = \frac{\chi_{n1}}{\omega C_n \chi_{n2}}.$$

Comparing this expression and (2.28), it is not difficult to make sure that when the condition (2.28) is satisfied, the values C_n are not negative. For exponential and linear distributions

$$C_{n3} = \frac{8\pi\varepsilon_0\chi_{n1}}{a^2 b\{1 + \coth[\alpha(n-1)b/2]\}}, C_{n2} = \frac{4\pi\varepsilon_0}{\alpha}\chi_{n1}(n-1), R_n - \frac{\chi_{n1}}{\omega C_n \chi_{n2}} . (2.36)$$

It is seen from (2.36) that in a particular case, if it is necessary to obtain a law of current distribution that is close to linear, the capacitances of loads should decrease and the resistances of the resistors should grow towards to the free end of antenna:

$$C_{n2} = C_{N2}(n-1)/(N-1), R_{n2} = R_{N2}(N-1)/(n-1), \qquad (2.37)$$

where C_{N2} and R_{N2} are the capacitance and the resistance of load located near the antenna base (near the generator). Thus, to create an in-phase current distribution that ensures high electrical characteristics of an antenna in a wide frequency range, each load should represent a parallel connection of a resistor and a capacitor. For the first time the expediency of using a complex load for creation of a linear current distribution was demonstrated in [10]. But later, the main attention was given to antennas with capacitive loads. As it follows from the given analysis and the calculations results, if resistors are included in parallel with the capacitors, then the linear law of current distribution along the radiator is observed more accurately and the operating frequency range increases. However, use of resistors leads to efficiency reduction, so the question of their application should be decided in each particular case.

If the loads are purely capacitive, then in accordance with (2.8) and (2.35)

$$\gamma_n^2 = -k^2 + \frac{k\chi_{n1}}{30b\omega C_n}. \qquad (2.38)$$

The propagation constant γ_n of the wave along this antenna is real at a given frequency f, if (2.27) does not violate, i.e., the capacitance of each load is not greater than

$$C_n \le \frac{\chi_{n1}}{30k^2 bc} = \frac{2.54 \cdot 10^5 \chi_{n1}}{f^2 b}. \qquad (2.39)$$

In Fig. 2.7b, the magnitude γ_n^2 is presented for an antenna with concentrated loads as a function of frequency. In the case of linear distribution of a current, capacitance C_{N2} of the load near the base of the asymmetric antenna is greater than the other capacitances and should be chosen in accordance with (2.39). Similarly, under other distributions, this expression determines the maximum capacitance.

Expression (2.38) determines the magnitude of the propagation constant along an in-phase antenna as a function of frequency. At low frequencies, γ_n^2 is real along the entire antenna. As the frequency increases, the values γ_n^2 gradually becomes purely imaginary, firstly on the sections adjoining to the generator, i.e., the current distribution along these sections of the radiator becomes sinusoidal, and the main lobe of the vertical directional pattern deviates from the horizontal direction. This effect limits the antenna frequency range from above (in Fig. 2.7b the corresponding point is indicated as f_{max}). From below, the range is usually limited by frequencies where the reactive component of input impedance is still too high.

In order that the magnitudes γ_n do not become purely imaginary as the frequency increases, the capacitances of loads must decrease at high frequencies, for example, must switch. In this case, as shown for the second option of fulfilling the condition (2.8), the capacitance should change in inverse proportion to the square of the average frequency. As one can easily verify, inequality (2.27) will be true at all frequencies, and constraint (2.29) will be removed, if the 'negative inductance' (2.34) will be connected in series with load Z_n from (2.36).

The analysis allows making a number of practical conclusions. In order that the concentrated loads may efficiently influence the current distribution, the distance between them must be small in comparison with the wavelength. For creating a wide-range radiator, only capacitors should be used as reactive elements, since inclusion of reactive two-poles of a more complex type, whose structure includes inductance coils, results in narrowing of the operating range. Capacitors enable creation, along a radiator in a wide frequency range, an electromagnetic wave with real propagation constant that corresponds to the exponential change of the current amplitude with positive decrement (the concave curve of the current). Obtaining a convex curve of a current with the help of simple concentrated elements (resistors, capacitors, inductance coils) is impossible.

The antenna with current distribution, close to linear, with capacitances decreasing in proportion to the distance from the free end of the arm has among distributions with positive α a higher matching level and a vertical directional pattern with narrow main lobe.

The described method of analysis on the one hand has sufficiently general character. It allows deriving of analytical expressions for impedances of loads, which ensure different laws of the current distribution along the radiator. On the other hand, it has approximate character, i.e., verification of the obtained results is necessary.

The method has been thoroughly tested by means of calculations and experiments [15]. The calculations were performed using a universal program based on the method of moments. The experimental results were obtained on a full-scale model. As an example, the input impedance of an asymmetric radiator, 6 m high and 13 mm in diameter, was calculated and measured. Along this radiator 10 capacitors were included at a distance 0.6 m from each other. The distance from the upper load to the antenna end and from the bottom load to its base is half as much. The capacitance of the bottom load was adopted equal to $18pF$, in which case the propagation constant γ_n is real along the entire antenna up to frequency of 40 MHz. The capacitances of the remaining loads are selected in accordance with (2.37).

Figure 2.8a shows curves for the active and reactive components of the antenna input impedance calculated depending on frequency as well as the curve for traveling wave ratio (*TWR*) in a cable with wave impedance of 75 Ohm. Here, for the sake of comparison, experimental values are also given. Figure 2.8b presents the calculated directional patterns in a vertical plane.

Calculation and experiment confirm that characteristics of antennas with loads are much better than their characteristics without loads. Under

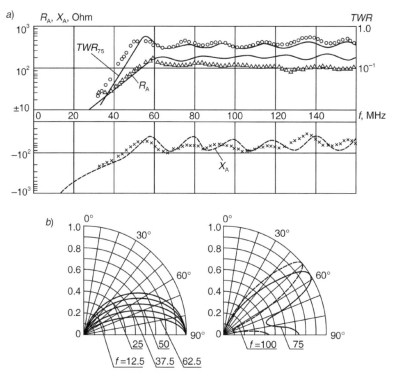

Fig. 2.8: Input characteristics (a) and vertical directional patterns (b) of a radiator with constant capacitive loads.

identical dimensions of antennas and identical requirements to their electrical characteristics, a frequency ratio (an overlap factor) of thin whip antennas is changed from 1.3 to 1.5, for antennas with capacitive loads ranging from 1.5 to 3.0, and for antennas with capacitors and resistors, from 3 to 4. The upper limit of a frequency ratio for antennas with loads is determined by a deviation of the main lobe of a vertical direction pattern from a horizontal direction. As for the impedance matching level, it remains high in a much wider range (with a frequency ratio of the order 10). Placement of inductance in the base of antenna with capacitors and resistors and selection of a cable facilitate a higher level of matching. Curves TWR_{200} and TWR^c_{200} demonstrate the level of matching with a cable, whose wave impedance is equal to 200 Ohm, without compensation of antenna reactance and with its compensation by means of the inductance 76.4 nH.

Calculations and experiment confirm also that antennas with frequency-dependent capacitances have a wider frequency range than the antennas with constant capacitances. Figure 2.9a shows TWR for three variants of monopoles excited by a cable with the wave impedance of 75 Ohm. The calculation was performed for the antenna which was 12 m high and 0.06 m in diameter with 10 capacitors located at a distance 1.2 m from each other (the upper and bottom capacitors are placed 0.6 m away from the monopole ends). Curve 1 in Fig. 2.9a corresponds to the radiator without loads, i.e., to the whip antenna, curve 2—to the radiator with loads, whose capacitances are frequency independent. Here the capacitance of the bottom load was chosen equal to 177 pF; in this case, the propagation constant is real along the entire antenna up to frequency 10 MHz. Capacitances of other loads are decreased in the direction to the free end of the antenna in proportion to the

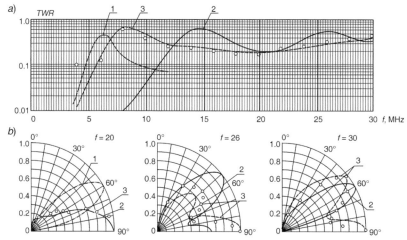

Fig. 2.9: *TWR* (a) and vertical directional patterns (b) of radiators without loads (1), with constant (2) and frequency-dependent (3) capacitive loads.

Table 2.1: Frequency ratio of radiators.

Version of antenna	Frequency, MHz		Range width Δf, MHz	Frequency ratio k_f
	f_1	f_2		
1	5.2	7.7	2.5	1.5
2	12.3	26.0	13.7	2.1
3	6.3	34.0	27.7	5.4

distance from it. This allows achievement the current distribution along the radiator, which is close to the linear law. Curve 3 is given for radiator with frequency-dependent capacitive loads. Their capacitances are changed in accordance with (2.34), where $f_0 = 20$ MHz.

Table 2.1 shows lower f_1 and upper f_2 frequencies of the operating range of each antenna. At a frequency f_1 TWR becomes greater than 0.2; at frequency f_2 a field strength along a perpendicular direction to an antenna axis becomes less than 0.7 of the maximum field strength (as a rule, f_2 corresponds to the second maximum on curve of TWR). TWR of the whip antenna with grow of frequency quickly decreases below level 0.2, and the frequency, corresponding to this point, is taken as f_2. Besides, Table 2.1 reports a range width $\Delta f = f_2 - f_1$ and overlap factor $k_f = f_2/f_1$.

As is seen from Fig. 2.9a and Table 2.1, the matching level of the variant 3 at low frequencies approaches the matching level of the whip antenna, and at high frequencies, the upper boundary of the operating range for variant 3 is shifted to the right in comparison with variant 2, since the main lobe of the vertical directional pattern deviates from the perpendicular direction to the dipole axis at a higher frequency. In addition, the minimum TWR in the middle of the operating range increases.

Figure 2.9a presents also the results of experimental verification of TWR for variant 3. The measurements were performed by means of an antenna model in a scale 1:10. The frequency range was split into short intervals and the capacitances of loads used in each interval were equal to capacitances calculated for the middle of the interval. The calculated and experimental results are in agreement with each other. Calculated curves and measured directional patterns in the vertical plane are demonstrated in Fig. 2.9b for the same variants of antennas.

The most practical and accessible realization of antennas with frequency-dependent capacitances is the use of tunable, in particular simply switched capacitors; for example, on the signal from the console (control panel).

2.3 Creating in-phase current using the method of a metallic long line with loads

Together with the method of impedance long line it was proposed another approximate method for calculating magnitudes of the loads, which provide the given current distribution [16]. This method allows to use of the given distribution in order to find the law, in accordance with which the equivalent length of a metallic long line grows along this line. This permits calculation of the load of each section corresponding to this law. This method and the method of impedance long line give analogous results.

Figure 2.3a shows an asymmetrical radiator of height L with N loads, which are uniformly located along the radiator at a distance b from each other. A current distribution in a first approximation is similar to a current distribution along a long line open at the end with the impedances Z_n connected in series. The distribution of a current along each section, located between neighboring loads, has sinusoidal character:

$$J(z_n) = J_n \sin k(z_n + l_{e,n-1}), \; 0 \le z_n \le b, \tag{2.40}$$

where $z_n = (N - n + 1)b - z$ is the coordinate measured from the end of section n; J_n is the amplitude of the current in section n; $l_{e,n-1}$ is the equivalent length of all preceding sections with $(n - 1)$ loads, whose total length is equal to $(n - 1)b$. The values l_{en} and $l_{e,n-1}$ are mutually related by the expression:

$$-jW \cot kl_{en} = Z_n - jW \cot k (b + l_{e,n-1}), \tag{2.41}$$

where W is the wave impedance of the line. Note that the value l_{en}, if $n = N$, is equal to the equivalent length L_e of the radiator.
Expression (2.41) permits to find the magnitude Z_n:

$$Z_n = -jW [\cot kl_{en} - \cot k(b + l_{e,n-1})]. \tag{2.42}$$

If the distance between loads is small ($kb \ll 1$), then

$$Z_n = -jW \frac{\sin k(b + l_{e,n-1} - l_{en})}{\sin kl_{en} \sin k(b + l_{e,n-1})} \approx -jW \frac{k(b + l_{e,n-1} - l_{en})}{\sin^2 kl_{en}}. \tag{2.43}$$

By means of expressions (2.42) and (2.43), one can calculate the magnitudes of loads. For this it is necessary to know the function in accordance with which the equivalent length grows along the line. The choice of this function depends on the current distribution along the radiator, which must be obtained in the first stage of solving the synthesis problem. In the general case

$$J(s_n) = J_A f(z), \ 0 \le s_n \le b, \ (N-n) \ b \le z \le (N-n+1)b, \tag{2.44}$$

where J_A is the current amplitude in the antenna base and $f(z)$ is the function of the current distribution.

Henceforth we shall assume that the function $f(z)$ is real and positive, i.e., we shall consider only in-phase distributions.

Suppose we want to obtain a given current distribution $J(z)$ along the antenna. For this we assume that the current $J(s_n)$ at the beginning and the end of each section coincides with the current $J(z)$. If the section's lengths are small, the current distribution along the line is close to the required one. In a general case, we have in accordance with (2.40) and (2.44)

$$J_{n+1} \sin k(b + l_{en}) = J_A f[(N-n-1)b] \text{ at } s_{n+1} = b,$$

$$J_{n+1} \sin k l_{en} = J_A f[(N-n)b] \text{ at } s_{n+1} = 0$$

If to divide the left and right parts of the first equality on to the corresponding parts of the second and to retain only the first terms of the expansion of trigonometric functions into a series, considering that the magnitude b is small, we will obtain $1 + kb \cot kl_{en} = f[(N-n-1)b]/f[(N-n)b]$, i.e.,

$$l_{en} = \frac{1}{k} \tan^{-1} \frac{kb}{f[(N-n-1)b]/f[(N-n)b]-1}. \tag{2.45}$$

As is seen from (2.45), the equivalent length l_{en} is frequency dependent. Knowing l_{en} and $l_{e,n-1}$, one may in accordance with the expression (2.4), find the loads magnitudes. They also have a frequency-dependent nature:

$$Z_n = j \frac{W}{kb} \left\{ \frac{f[(N-n-1)b]}{f[(N-n)b]} - 2 + \left(1 + k^2 b^2\right) \frac{f[(N-n+1)b]}{f[(N-n)b]} \right\}. \tag{2.46}$$

An input impedance of an open at the end long line is a first approximation to a reactive component of antenna input impedance. In a general case, it is equal to

$$Z_A = -jW \cot kL_e = -j \frac{W}{kb} [f(-b) - 1]. \tag{2.47}$$

This expression shows that the function $f(z)$ should change slowly near the antenna base and the difference $f(-b) - 1$ should be small, of the order of kb. Otherwise, the reactive component of an input impedance will be great.

The linear distribution (2.15) is a particular case of exponential distribution (2.16). The equivalent lengths of the long lines for exponential and linear distribution in accordance with (2.45), (2.16) and (2.15) are equal to

$$l_{en3,4} = \frac{1}{k}\tan^{-1}\frac{kb[1 - \exp(-n\alpha b)]}{\exp(\alpha b) - 1}, \; l_{en2} = \frac{1}{k}\tan^{-1}(nkb). \tag{2.48}$$

As is seen from (2.48), if $\alpha > 0$, the equivalent length of the antenna arm may not exceed quarter of the wavelength. The input impedance of a long line with exponential and linear current distribution may be written in the form

$$Z_{A3,4} = -j\frac{W[\exp(\alpha b) - 1]}{kb[1 - \exp(-\alpha L)]}, \; Z_{A2} = -jW/(kL). \tag{2.49}$$

Using expressions (2.46) and (2.24), we find the magnitudes of the loads, which provide the exponential law of current amplitude distribution along the radiator:

$$Z_n = -j\frac{W}{kb}\frac{2(ch\alpha b - 1) + k^2 b^2 \exp(-\alpha b)}{1 - \exp(-n\alpha b)}\{1 - \exp[-(n-1)\alpha b]\}. \tag{2.50}$$

If the product αb is not small, then neglecting by second term of the numerator, we obtain

$$Z_n = 1/(j\omega C_n), \tag{2.51}$$

where $C_n = \dfrac{b[1 - \exp(-n\alpha b)]}{2Wc(\cosh \alpha b - 1)}$, and as is seen from this formula, sign of C_n coincides with sign of α.

Thus, in order to obtain an exponential distribution of current amplitude with a decrement of enough great magnitude, one must use capacitive loads. Capacitors allow creation of only a concave current distribution (with $\alpha > 0$). In order to obtain a convex distribution when $\alpha < 0$, capacitances must be negative. If $|\alpha|b \ll 1$, then, confining by the first terms of the functions expansion into a series, we find from (2.50)

$$Z_n = 1/(j\omega C_n) + j\omega \Lambda_n, \tag{2.52}$$

where

$$C_n = n/(cW\alpha), \quad \Lambda_n = -Wb(n-1)/(cn).$$

In order to obtain the exponential distribution with a small decrement α using capacitors, the negative inductances Λ_n should be included. They can be neglected, if the first term of (2.50) is much larger than the second one, i.e., $\alpha \gg k^2 b(n-1)$. If $\alpha = 0$,

$$Z_n = j\omega\Lambda_n = -jkbW\,(n-1)/n, \tag{2.53}$$

i.e., in case when the loads are made in the form of negative inductances, proportional to $(n-1)/n$, we obtain a purely linear distribution.

As it follows from the made of this section analysis, the method of a metallic long line with loads and the method of the impedance long line lead to similar results. Comparison of these results allows application of these methods to specific problems, using specific details of the current distribution along the radiators. Results, obtained by means of these methods, can be used for solving an optimization problem of the antenna with loads by the mathematical programming method.

2.4 Optimization of antenna characteristics by means of a mathematical programming method

Use of the mathematical programming methods [17] plays a major role in solving the inverse problems of antenna engineering. These methods allow determination of optimal parameters of an antenna, in particular selection of its geometric dimensions and magnitudes of loads for creating antennas with specified characteristics, or more precisely, with characteristics that are as close as possible to given ones.

This remark is due to the fact that the variation interval of radiator parameters is bounded, i.e., not every value of antenna electrical characteristic can be realized practically. Different characteristics are optimal for different parameters. Moreover, an antenna should exhibit certain properties not at a single fixed frequency, but in the entire operation range. Therefore, the selected parameters are a result of a compromise reached with the help of the mathematical programming method.

The problem of mathematical programming in the general case is stated as follows: It is necessary to find vector $\vec{x}$ of parameters that minimizes some objective function $\Phi(\vec{x})$ under imposed constraints $\Phi_i(\vec{x}) \geq 0$. Depending on the type of functions $\Phi(\vec{x})$ and $\Phi_i(\vec{x}) \geq 0$ mathematical programming is divided into linear, convex and non-linear ones. In the case at hand, the problem is solved by non-linear programming methods since the type of function $\Phi(\vec{x})$ is unknown.

The objective function $\Phi(\vec{x})$ (or general functional) is a sum of several partial functional $\Phi_j(\vec{x})$ with weighting coefficients p_j and penalty functions Φ_{ij}:

$$\Phi(\vec{x}) = \sum_j p_j\, \Phi_j(\vec{x}) + \sum_i \Phi_{ij}. \tag{2.54}$$

The partial functional is an error function for one or the other antenna characteristic. The weighting function allows taking into account the importance of this characteristic and the sensitivity of corresponding functional to changes of vector $\vec{x}$. A penalty function is zero if the parameters lie within a given interval and has great value even if only one of the parameters falls outside the interval limits. For an antenna with concentrated loads the controlled parameters x are magnitudes of loads, coordinates z_n of their connection points and the wave impedance W of a cable. Under loads understood simple elements: capacitors with capacitance C_n, coils with inductance Λ_n and resistors with resistance R_n. Values z_n, W, C_n, Λ_n and R_n are to be real, positive and frequency-independent, and z_n are to be smaller than antenna length L. These requirements naturally limit the variation interval of the parameters.

Different ways of an error function formation are known. Good results are produced by means of quasi-Tchebyscheff criterion:

$$\Phi_j(\vec{x}) = \frac{1}{N_f}\left[\frac{f_{j0}}{f_{j\min}(\vec{x})} - 1\right]\left\{\sum_{n_f}\left[\frac{f_{j0}/f_j(\vec{x}) - 1}{f_{j0}/f_{j\min}(\vec{x}) - 1}\right]^S\right\}^{1/S}, \tag{2.55}$$

where N_f is a number of points of the independent argument (e.g., a number of frequencies in a given range), n_f is a frequency number, $f_j(\vec{x})$ is one of the electrical characteristics of an antenna, $f_{j\min}(\vec{x})$ is its minimal value in the considered interval, f_{j0} is a hypothetical value of the characteristic, which must be reached, S is the index of power, allowing control of the method sensitivity.

Another criterion is called a root-mean-square criterion. It uses another error function:

$$\Phi_j(\vec{x}) = \frac{1}{N_f N_l}\sum_{n_f=1}^{N_f}\sum_{n_l=1}^{N_l}\left[f_j(\vec{x}) - f_{j0}\right]^2, \tag{2.56}$$

where N_f and N_l are numbers of points of an independent argument (e.g., a number of frequencies in a given range and a number of division points on the wire), n_f is a frequency number, n_l is a point number, $f_j(\vec{x})$ is one of the antenna electrical characteristics (e.g., a current or a voltage), f_{j0} is a hypothetical value of the characteristic, which must be reached.

The choice of optimizable characteristics depends on a stated problem. For creation of a wide-band radiator one must use as characteristics $f_j(\vec{x})$, a travelling wave ratio (*TWR*) in a cable and a pattern factor (*PF*), which is equal to the average level of radiation at predetermined angles range. If resistors with resistances R_n are used as loads, it is necessary to supplement the set of $f_j(\vec{x})$ by the characteristic of antenna efficiency (η_A):

$$TWR = \frac{2a}{a^2 + b^2 + 1 + \sqrt{\left(a^2 + b^2 + 1\right)^2 - 4a^2}}, PF = \frac{1}{K}\sum_{k=1}^{K}F(\theta_k), \eta_A = 1 - \frac{1}{J_A^2 R_A}\sum_{n=1}^{N}|J_n|^2 R_n,$$

(2.57)

where $a = R_A/W, b = X_A/W$ are respectively ratios of active and reactive components of antenna impedance to a wave impedance of a cable, K is number of angles θ_k in a vertical plane within the limits of an angular sector from θ_1 to θ_k (for example, from 60° to 90°), k is an angle number within this sector, and $F(\theta_k)$ is a magnitude of normalized directional pattern in the vertical plane for an angle θ_k. N is the number of loads, J_n and J_A are the currents in the load n and in the antenna base, respectively.

If it is necessary to obtain a given current distribution $J(z)$, it is expedient to use as characteristic $f_j(\vec{x})$ either real and imaginary current components:

$$f_1 = \text{Re } J(z, f), f_2 = \text{Im } J(z, f),$$

(2.58)

or amplitude and phase of a current:

$$f_3 = |J(z, f)|, f_4 = \tan^{-1}[\text{Im}(z, f)/\text{Re}(z, f)].$$

(2.59)

In cases when the analytical expression for objective function $\Phi(\vec{x})$ is absent, the minimum of function must be found by a numerical method based on searching the gradient. The gradient searching is an iterative procedure, which becomes step by step from one set of parameters $\vec{x}_M$ to another set $\vec{x}_{M+1}$ in a direction of maximal decrease of the objective function. Therefore, this method is called the method of steepest descent:

$$\vec{x}_{M+1} = \vec{x}_M - \alpha_M \, grad \, \Phi(\vec{x}_M),$$

(2.60)

where M is the iteration number, α_M is the scale parameter. Each iteration in essence searches the minimum of a functional (of an objective function) in the direction of anti-gradient. As a result of this search, the coefficient $\alpha_{M'}$ at which the objective function $\Phi(\vec{x}_M)$ entering into equation (2.60) becomes minimal, is calculated. Also, the values of parameters $\vec{x}$, which correspond to this minimum are determined.

A modification of the steepest descent method is of the conjugate gradients. In this case the iteration 1, (Q-1), (2Q+1) and so on are calculated according to the anti-gradient (here Q is number of parameters) and the rest of the steps correspond to the expression

$$\vec{x}_{M+1} = \vec{x}_M - \alpha_M \vec{G}_M, \tag{2.61}$$

where

$$\vec{G}_M = grad\Phi(\vec{x}_M) + \left| \frac{grad\Phi(\vec{x}_M)}{grad\Phi(\vec{x}_{M-1})} \right|^2 \vec{G}_{M-1}.$$

The calculation ends when the decrease of the objective function from iteration to iteration becomes smaller than a preset value, or the number M of iterations exceeds certain limit M_0.

The minimum of an objective function $\Phi(\vec{x}_M)$ and the values of parameters, which correspond to this minimum, are determined in each iteration. In essence, each iteration searches a parameter α_M. The most rational method consists in a sequential increase (for example, doubling) of magnitude α and further interpolating the function $\Phi(\vec{x}_M)$ in a considered interval by a polynomial of a given power. It is convenient to use cubic interpolation since the number of interpolation nodes is large enough (four), and the root of the derivative (the value α that turns the derivative into zero) is found analytically. If the first step results in an increase, rather than decrease of the objective function, the step should be reduced by a factor of 10^p, where $p = 1, 2...,$ whereupon the linear search goes on again with doubling of a step.

The mathematical programming method (synthesis) presupposes frequentative computation of the antenna's electrical characteristics at different initial parameters (analysis). Performing such calculations requires incorporation of a special program into the synthesis software. This program allows determining of all electrical characteristics of an antenna, i.e., calculating the functions $f_i(\vec{x})$ for vector $\vec{x}$ of initial parameters at given emfs and loads. The most laborious of these calculations is computation of self and mutual impedances between antenna sections (between so-called short dipoles). Therefore, in order to speed up the calculations, it is expedient to fixate, for example, points of placing concentrated loads in order that the coordinates of short dipoles and their mutual impedances do not change from iteration to iteration. If there are enough many loads, i.e., the distances between them are small in comparison with the wavelength, this restriction will have no effect on the synthesis results.

As regards initial magnitudes of concentrated loads, these magnitudes must be found by the approximate physical method described in Sections 2.2 and 2.3. The results of the calculations show that the computational process in this case speeds up. But the most important result of using these magnitudes consists in reducing the error probability since an arbitrary choice of the initial parameters may cause the optimization process, which will lead to a local, rather than true, minimum of the objective function.

The described iterative procedure is essentially motion in the vector space $\vec{x}_m$ along the surface $\Phi(\vec{x}_M)$ in the direction of a minimum of a general functional. A particular case of such motion, when the vector $\vec{x}_m$ consists of two parameters (x and y) is shown in Fig. 2.10. The algorithm of the iterative procedure is presented in Fig. 2.11.

It is necessary to emphasize that the method of mathematical programming makes it possible to realize the synthesis of a wide-range radiator, i.e., choose the optimum capacitive loads that provide the highest possible level of *TWR, PF* and efficiency in a given frequency range. The synthesis program, using the mathematical programming method, helps to bring the problem solution to an end.

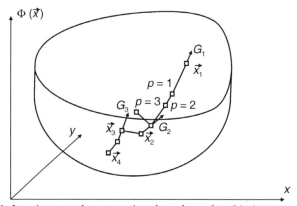

Fig. 2.10: Iterative procedure as motion along the surface $\Phi(x_M)$ to an optimum.

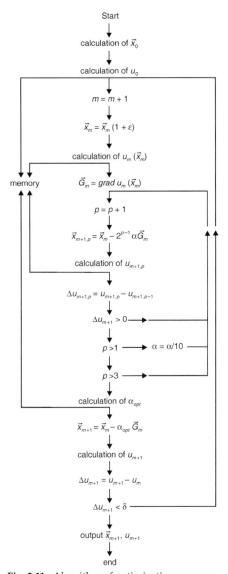

Fig. 2.11: Algorithm of optimization program.

3

Solved Problems

3.1 Optimal matching of linear radiators with constant capacitive loads

In Chapter 2 the problem of creating a wide-range radiator based on employment of capacitive loads was considered. This radiator provides both high level of matching with a cable and a desired shape of the directional pattern in a vertical plane. The method of impedance line described in this chapter gives an approximate solution that allows use of mathematical programming for finding antenna loads. Characteristics of the antenna with obtained loads must be maximally closest to the required one. The application of this approach confirms its validity and usefulness.

The program of synthesis was used for selection of the optimal capacitive loads, allowing achievement of maximal TWR and PF in the predetermined range of frequencies $f_1 - f_2$. All weighting coefficients p_j as a rule were taken by identical. The calculations showed that for a synthesis of a simple antenna, four to five iterations are enough. The number of optimizable electrical characteristics in this case has little effect on the synthesis time. For example, the time of optimizing TWR and PF is almost equal to the time of optimizing only TWR. The choice of criteria has practically no effect on the results of synthesis wide-range radiators and the calculation time. In particular, the root-mean-square criterion has no advantage as compared with the quasi-Tchebyscheff criterion and is rarely used. As a hypothetical value f_{j0} of the characteristic, it is advisable to choose a maximum magnitude, since its decrease leads to reduction of the result. An increase of index S in (2.54) accelerates the process convergence. In the calculations it was assumed that $S = 6$.

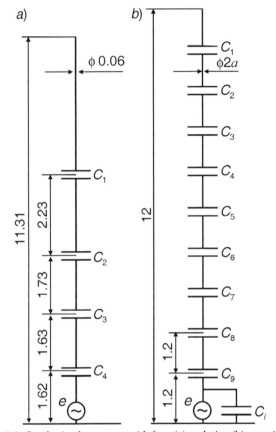

Fig. 3.1: Synthesized antennas with four (a) and nine (b) capacitors.

Figure 3.1 gives the basic dimensions of two optimizable antennas with loads. The first antenna is the monopole of height L = 11.31 m with four capacitors, irregularly spaced along the radiator. The second antenna is the monopole of height L = 12 m with nine capacitors spaced equidistantly. The capacitance C_i, equal to capacitance of a typical ceramic insulator (15 pF), is located at the base of the second antenna in parallel with its input.

The results of the antennas synthesis are presented in Tables 3.1 and 3.2. The basic characteristics of the radiators are given in Table 3.1, where the following designations are used: a is antenna radius, N is number of capacitors, N_f is number of used frequencies, M_0 is a required number of iterations. *TWR* and *PF* of the first antenna variant as functions of a frequency for different iterations are shown in Figs. 3.2a and 3.2b. Here M is the iteration number. The input characteristics and the directional patterns of the synthesized antenna are presented in Figs. 3.2c and 3.2d respectively. As is seen from the figures, the curve of *TWR* has a maximum

Table 3.1: Main characteristics of antenna.

Version	L, m	a, m	N	$f_1 - f_2$, MHz	N_f	M_0	TWR_{min}	PF_{min}
1	11.31	0.03	4	11.5–16.5	11	4	0.310	0.860
2	"	"	"	"	"	5	0.360	0.812
3	12	0.03	9	7.5–15	16	4	0.123	0.819
4	"	"	"	15–30	"	4	0.273	0.610
5	"	"	"	30–60	"	5	0.360	0.562
6	"	0.15	"	7.5–15	"	4	0.205	0.813
7	"	"	"	15–30	"	5	0.414	0.680
8	"	"	"	30–60	"	4	0.380	0.605
9	"	0.03	"	8.5–13	10	3	0.217	0.870
10	"	"	"	13–22	"	5	0.216	0.790
11	"	"	"	22–60	20	8	0.204	0.437
12	"	0.15	"	8.5–13	10	5	0.314	0.829
13	"	"	"	13–22	"	4	0.278	0.859
14	"	"	"	22–60	20	5	0.322	0.565

Table 3.2: Optimal capacitive loads.

Version	Optimal capacitive loads, pF								
	C_1	C_2	C_3	C_4	C_5	C_6	C_7	C_8	C_9
1	44.3	33.2	91.2	432	-	-	-	-	-
2	84	164	143	182	-	-	-	-	-
3	37.3	81.1	127	181	2 58	369	516	6 91	883
4	8.7	20.6	36.5	51.1	58.7	53.3	50.6	88.1	156
5	2.0	3.9	5.4	6.1	9.5	12.2	15.7	21.1	18.3
6	51.2	134	219	340	477	633	804	981	1150
7	10.7	28.4	57.0	76.7	86.4	86.8	53.5	216	409
8	4.5	19.0	11.4	15.8	26.4	29.8	32.4	23.0	35.7
9	33.9	72.8	115	164	223	296	385	492	608
10	8.4	18.4	30.3	40.4	47.1	53.3	82.8	151	248
11	1.7	5.6	12.2	11.7	17.6	41.0	30.5	56.4	240
12	21.1	0.2	39.2	231	519	909	1380	1900	2450
13	20.3	76.3	122	78.5	0.1	107	351	761	1340
14	2.9	26.9	0.3	42.1	22.6	59.1	35.8	55.0	73.1

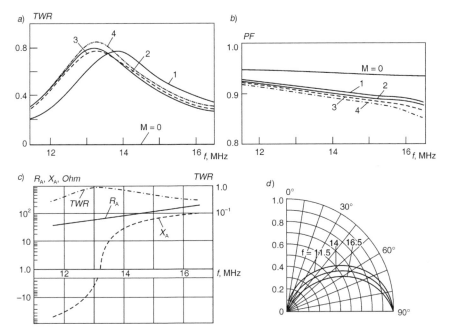

Fig. 3.2: Change of *TWR* (a) and *PF* (b) depending on iteration number; the input impedance (c) and the directional pattern (d) of the synthesized antenna with the height 11.31 m.

in an area of first (series) resonance, and a magnitude of *PF* decreases with frequency growth. The capacitances increase towards the antenna base, but not monotonically. Optimal capacitive loads are given in Table 3.2.

The second option of optimization is distinguished from the first option by the fact that the wave impedance of a cable is used as a regulated parameter in addition to the magnitudes of the capacitances. The optimal wave impedance is $W = 238$ Ohm. In this case the antenna has optimal characteristics in an area of second (parallel) resonance.

The other variants of Tables 3.1 and 3.2 concern the second antenna with an insulator at the base. Frequency ratio for the antenna variants with numbers 3–8 is adopted as equal to two. As seen from Table 3.1, the increase of antenna radius from 0.03 m to 0.15 m at frequencies up to 30 MHz results in growing minimal *TWR*, approximately by a factor 1.5. The variation of radius has a weaker influence on *PF* when it is minimal.

Figure 3.3 shows the electrical characteristics of variants 3–5 (antenna radius is 0.03 m) and of a whip antenna with the same geometrical dimensions and the same capacitance C_i of the insulator. The characteristics

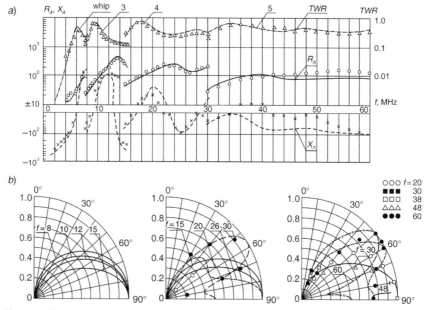

Fig. 3.3: The input characteristics of 12-meter antennas of radius 0.03 m (a) and their directional patterns in a vertical plane (b).

of the antenna with a radius 0.15 m (variants 6–8) are similar. As seen from the calculations, the curve of *TWR* can have two maximums at high frequencies. The curves of *PF* do not decrease monotonically with frequency but have maximums too. In addition to calculated curves, the pictures demonstrate (by dots and other symbols) the results of experimental verification, carried out on mock-ups in scale 1:5. The calculation and experiment coincide well.

As it follows from Table 3.1 (variants 3–8), if frequency ratios of various ranges are identical, the level of antenna matching with a cable is different for the different ranges. This level substantially rises if the frequency grows. In order to obtain more uniform and, on the whole, better characteristics over the entire frequency range (at unchanged number of sub ranges), it is expedient to split the total range into such parts that the frequency ratio of sub ranges increases with increase in frequencies. The results of solving this problem are presented in Table 3.1 as variants 9–14.

The electrical characteristics of variants 12–14 of radius 0.15 m as well as of a monopole without loads of the same radius and with capacitance C_i at the base, are given in Fig. 3.4. Data of Table 3.1 confirm a general increase of *TWR* level in comparison with variants 3–8. In all the sub ranges, increase in the antenna radius causes a rise of minimum *TWR* (approximately by a factor 1.5), together with a rise of minimum *PF* at high frequencies.

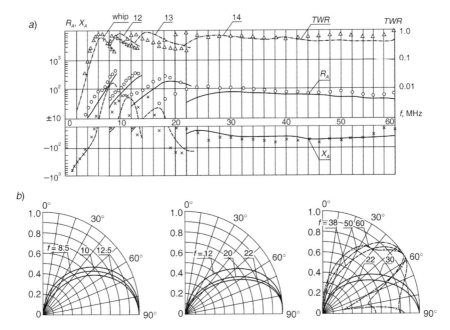

Fig. 3.4: Input characteristics of 12-meter antennas of radius 0.15 m (a) and their directional patterns in a vertical plane (b).

The results of optimization of 12-meter antennas with capacitances $C_i = 15$ pF at the base are used to plot in Fig. 3.5 with the curves for the minimal *TWR* depending on relative antenna length L/λ_{max} (λ_{max} is the maximum wavelength of the range) at various frequency ratios k_f and different antenna radii a. These curves determine the maximum attainable characteristics, which can be obtained with the help of antennas with constant capacitive loads.

Results of calculations and experimental verification presented in this section show that the described procedure is an effective way to optimize the characteristics of antennas with capacitive loads. Application of this procedure confirms its validity and usefulness.

The calculation results also show that if it is necessary, the antenna range can be expanded in the direction of high frequencies at a sufficiently high *TWR*; for example, one can obtain *TWR* ≥ 0.2 in the range with a frequency ratio of about 10. But the directional pattern in the additional (high-frequency) range deteriorates substantially. In this connection, the frequency ratio of an antenna with capacitive loads does not exceed 3 (at *PF* ≥ 0.5 and *TWR* ≥ 0.2). As it becomes clear later, this circumstance was caused by the fact that under calculations of the objective function $\Phi(\vec{x})$ in accordance with (2.53) weight coefficients p_j were adopted, equal

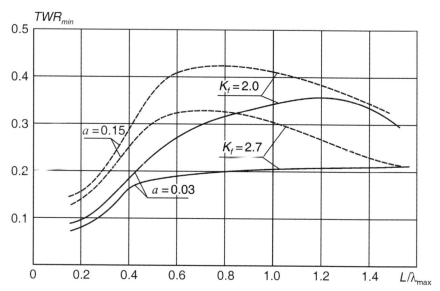

Fig. 3.5: The maximum level of matching for the antenna with constant capacitances.

in magnitude, without taking into account the different sensitivity of the partial functional $\Phi_j(\vec{x})$ to the change of the vector $\vec{x}$.

3.2 Creating a required current distribution in a given frequency range

The procedure developed for solving the described problem can be effectively used to solve other problems. They are discussed in other sections of this chapter. Let's start with creating a desired current distribution in a certain frequency range. It should be emphasized that the requirement to create a specific distribution does not mean a strict coincidence at all frequencies of the given and the obtained distribution. This requirement means only obtaining a distribution which is close to the required as far as possible.

Synthesis of antennas with a given current distribution in a certain frequency range was considered at examples in [16]. The calculation was performed for the monopole of a height 6 m with 10 capacitive loads, described in Section 2.2. Initial calculation of loads magnitudes was executed by means of approximate method of a metallic long line with loads. Figure 3.6a shows the equivalent lengths l_{en} of this line measured from the free end of monopole to the points n of capacitors location. The capacitances C_n of these loads are given in Fig. 3.6b.

In accordance with the two possible tasks, equivalent lengths and capacitances were calculated at a frequency of 40 MHz for two variants of

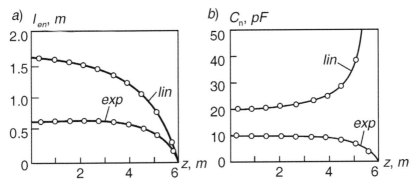

Fig. 3.6: Results of approximate calculation of equivalent lengths (a) and capacitances (b).

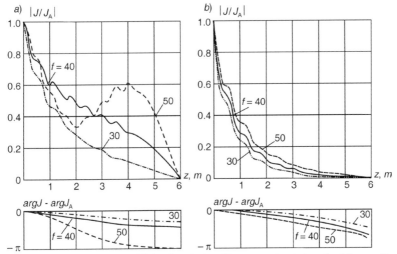

Fig. 3.7: Currents in antennas with loads, calculated by approximated method for creating at $f = 40$ MHz linear (a) and exponential (b) distribution of the amplitude.

current distributions: linear and exponential (with a logarithmic decrement $\alpha = 2$). The corresponding curves are indicated by labels 'lin' and 'exp'. Equivalent lengths and capacitances are calculated by means of (2.48) and (2.52) respectively.

These results were used for rigorous calculation of current amplitude and phase along antennas. They are given in Fig. 3.7a for a linear distribution and in Fig. 3.7b for an exponential distribution. As can be seen from the figures, at $f = 40$ MHz the amplitude distribution is close to the required one and the phase curves have a slight slope. When the frequency changes, i.e., at $f = 30$ and 50 MHz, the amplitude-phase distribution of the current is not conserved.

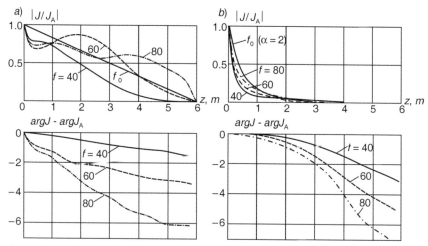

Fig. 3.8: Results of synthesis of linear (a) and exponential (b) current distribution in a range from 40 to 80 MHz.

In order to provide the required current distribution in a range of 40 to 80 MHz, the synthesis of antennas was performed by the method of mathematical programming. These results are shown in Fig. 3.8 for the linear (a) and exponential (b) distributions respectively. The currents amplitude and phase are obtained by means of optimizing antennas' electrical characteristics. The objective function was formed, using the root-mean-square criterion. Parameters calculated by the method of a metallic long line with loads at the middle frequency $f = 60$ MHz were taken as an initial approximation. The number N_f of used frequencies in a given range is equal to 9 and number N_l of division points on a wire is equal to 11.

The results were improved significantly. In each figure, four curves for current amplitude are given: the curve, labeled by f_0, corresponds to the required distribution and curves labeled by $f = 40$, 60 and 80 correspond to the synthesis result at frequencies 40, 60 and 80 MHz. As is seen from the figures, the obtained distribution is, on the whole, close to the required distribution, but is not identical to it. However, this difference is caused primarily by limited potential opportunities of the antennas. These limited opportunities rather than an inexact method are the reason for this difference. Thus, in addition to the successful solution of a problem, the used methods permit determination of the potential opportunities of antennas.

3.3 Reducing the influence of nearby metal superstructures

The loads included in the antenna allow attainment of the concrete electrical characteristics for a selected antenna length. If it is possible to manufacture and install antennas of the required length, then the freedom in choosing

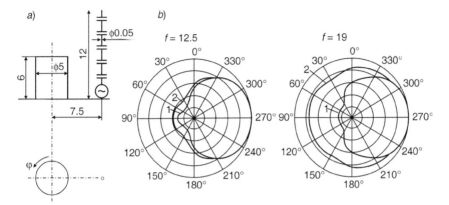

Fig. 3.9: An antenna near a superstructure (a) and its horizontal directional pattern (b).

this length permits weakened influence of closely located metal structures; for example, of superstructures, on the directional pattern of antenna and antenna array [18]. Figure 3.9 demonstrates results of calculating the directional patterns of a monopole situated near a metal superstructure in the shape of a circular metal cylinder of finite length. The directional patterns in the horizontal plane are presented at two frequencies of *HF* range.

Two variants of antennas are examined: (1) The monopole without loads of a height 6 m and of a diameter 0.016 m. (2) The radiator of a height 12 m and of a diameter 0.06 m with nine capacitive loads, ensuring optimal electrical characteristics in the frequencies ranging from 8 to 22 MHz. The relative arrangement of the superstructure and the antennas as well as the superstructure dimensions are given in Fig. 3.9a. The calculations were performed by means of a program based on the method of moments. The circular cylinder during calculation was replaced with a wire structure consisting of equidistant wires, located along generatrices of the cylinder and the radii of its upper surface. As is seen in Fig. 3.9b, the radiation of an ordinary monopole (curve 1) in the direction of superstructure decreases sharply, and the use of the antenna with loads (curve 2) allows weakening of this effect.

Figure 3.10 demonstrates similar results for a uniform linear array situated near the superstructure. The same two variants of antennas are adopted as the array elements. The mutual arrangement of the superstructure and the antennas as well as the superstructure dimensions are given in the figure and the phase shift between antenna currents is adopted as zero. The calculation results show that in an upper part of a frequency range the superstructure influence on the directional pattern of array, consisting of monopoles without loads, is slighter than its influence on the directional pattern of the sole monopole. This is apparently concerned with the fact that the superstructure does not hinder the propagation of electromagnetic

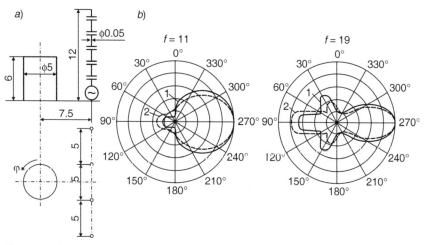

Fig. 3.10: A linear array near a superstructure (a) and its directional pattern in the horizontal plane (b).

waves from the side antennas. Nevertheless, the use of radiators with loads in this case allows reduction in the superstructure influence and increase in the signal in its direction.

3.4 An optimal matching of V-radiators with constant capacitive loads

Capacitive loads can also be used to improve the electrical characteristics of a directional V-antenna [19]. If the arm length L is larger than 0.7λ, the conventional V-antenna becomes directed with predominant radiation along the bisector of its aperture. However, with growing frequency the main lobe of the directional pattern in the vertical plane splits and the radiation along the bisector decreases. The inclusion of capacitive loads in antenna wires allows to extend the frequency range, in which directional radiation along the bisector is created, and to increase a useful signal in this direction.

A symmetric V-radiator with an arm length L and an arbitrary angular aperture $\alpha = \pi - 2\theta_0$ is presented in Fig. 3.11. The far field $E_\theta(0)d\varsigma$ along an angle bisector, created by an element $d\varsigma$ of the upper arm located near the origin of the coordinates, is equal to

$$E_\theta(\varsigma)d\varsigma = \frac{E_\theta(0)J(\varsigma)}{J(0)} \exp\left(jk\,\varsigma\sin\theta_0\right)d\varsigma, \tag{3.1}$$

where ς is the coordinate measured along the antenna arm, $J(\varsigma)$ is the current along the upper arm, located near point 0 with a current $J(0)$, and $k\varsigma \sin\theta_0$

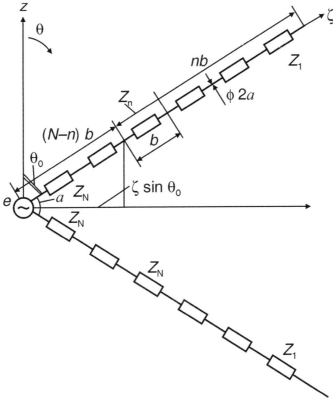

Fig. 3.11: V-dipole with loads.

is the path-length difference between the points 0 and ς. To ensure that all the far fields of different points coincide in phase, the current distribution along the arm must correspond to the expression

$$J(\varsigma) = J(0)\, f(\varsigma)\, \exp\left(- jk\, \varsigma \sin \theta_0\right). \tag{3.2}$$

Here $f(\varsigma)$ is a real and positive function.

Let loads Z_n are connected uniformly at a distance b from each other along a wire of each antenna arm. If the load spacing is small, then, as for the linear dipole, the replacement of concentrated loads by distributed surface impedance practically does not change the current distribution along the antenna. We assume that the surface impedance of each antenna section with load Z_n is constant and (2.7) is valid.

As is said in Section 2.1, the current distribution along the antenna with piecewise constant surface impedance coincides in the first approximation with current distribution along an equivalent impedance line, i.e., along the line with stepped variation of propagation constant (see Fig. 2.3). Here, the wave propagation constant γ_n along the section n is related to surface impedance $Z^{(n)}$ in accordance with (2.8). If the law of changing propagation constant is known, one can use magnitude γ_n for calculation with the help of (2.8) concentrated loads Z_n, which are needed for the embodiment of this law. The current along the section n of a stepped line is

$$J(\varsigma_n) = I_n sh(\gamma_n \varsigma_n + \varphi_n),\ 0 \le \varsigma_n \le b, \tag{3.3}$$

where I_n and φ_n are a current amplitude and phase at a section n, respectively, and ς_n is the coordinate, measured from the section end, i.e., $\varsigma_n = (N - n + 1)b - \varsigma$. We equate current $J(\varsigma_n)$ at the beginning and the end of each section to current $J(\varsigma)$ to ensure the phase coincidence of far fields from all antenna sections. The current inside each section does not coincide with current $J(\varsigma)$. However, if the sections' lengths are small, the current distribution along the line is close to $J(\varsigma)$.

According to (3.2) and (3.3), at $\varsigma_n = b$ and $\varsigma_n = 0$:

$$I_n sh(\gamma_n b + \varphi_n) = J(0) f [(N - n)b] \exp[-jk (N - n)b \sin \theta_0],$$

$$I_n sh\ \varphi_n = J(0) f [(N - n + 1)b] \exp[-jk (N - n + 1)b \sin \theta_0].$$

If to divide the left- and right-hand sides of the first equation on to the corresponding sides of the second equation, then considering that b is a small value and retaining only the first terms of series expansions for trigonometric functions of small arguments, we get similarly (2.11) for sections n and $(n + 1)$

$$\tanh \varphi_n = \gamma_n b \bigg/ \left\{ \frac{f[(N - n)b]}{f[(N - n + 1)b]} (1 + jkb \sin \theta_0) - 1 \right\}, \tag{3.4}$$

$$\tanh \varphi_{n+1} = \gamma_{n+1} b \bigg/ \left\{ \frac{f[(N - n - 1)b]}{f[(N - n)b]} (1 + jkb \sin \theta_0) - 1 \right\}. \tag{3.5}$$

Voltage and current are continuous along a stepped line. Therefore, (2.13) is true. Together with (3.4) and (3.5), it forms a set of equations that allows to relate γ_n and γ_{n+1}. From solution of this set of equations, it follows that magnitude γ_n is independent of γ_{n+1}:

$$\gamma_n = \frac{1}{b} \sqrt{1 - \frac{2f[(N - n)b] - f[(N - n - 1)b]}{f[(N - n + 1)b]} - 2jkb \sin \theta_0 \frac{f[(N - n)b] - [(N - n - 1)b]}{f[(N - n + 1)b]}}.$$
$$\tag{3.6}$$

This expression generalizes (2.14) and transforms into it at $\theta = 0$.

The possibility of implementation of propagation constant γ_n is determined by the possibility of implementation of concentrated loads. According to (2.8), at low frequencies, when inequality (2.29) and equality (2.30), which follows from (2.29), are true, the load value is

$$Z_n = -j30(\gamma_n b)^2/(kb\,\chi_n). \tag{3.7}$$

By substituting (3.6) into (3.7), we get

$$1/Z_n = j\omega C_n + 1/R_n, \tag{3.8}$$

where

$$R_n = \frac{60\sin\theta_0}{\chi_n}\frac{f[(N-n-1)b]-f[(N-n)b]}{f[(N-n+1)b]}, \quad C_n = 4\pi\varepsilon_0 b\chi_n\left\{1-\frac{2f[(N-n)b]-f[(N-n-1)b]}{f[(N-n+1)b]}\right\}^{-1}.$$

As seen from (3.8), each load should be a series connection of a resistor and a capacitor, where the resistance of the load is positive, if function $f(\varsigma)$ decreases monotonically with growing ς, and the capacitance of the load is positive, if function $f(\varsigma)$ is concave. Here, the resistance depends on the angular aperture of the antenna and the form of function $f(\varsigma)$, whereas the capacitance depends only on the latter.

For a linear radiator with loads ensuring the maximal radiation in the plane, perpendicular to the radiator's axis, each load should, when condition (2.8) holds, represent a capacitor. Capacitors ensure real propagation constant γ_n and an in-phase current distribution along an antenna. In V-antenna with capacitors the propagation constant is real and current is in-phase. The resistor must be located in series with the capacitor, and that leads to a phase lag of current along an antenna wire. Such phase delay is necessary for a V-dipole, since it compensates the path-length difference from individual dipole sections to an observation point and ensures coincidence of phases of fields created in the far zone by sections along the bisector of the angular aperture.

The use of resistors in a transmitting antenna is inexpedient. This means that the loads of a V-dipole should not differ from the loads of a linear radiator, which ensure an in-phase current distribution along an antenna. At high frequencies, when condition (2.8) does not perform, the current along a linear radiator is in-phase, if the load represents a negative inductance (a capacitance, which is inversely proportional to square of frequency). Similarly, the load for a V-dipole should be a series connection of a capacitor with a frequency-dependent capacitance and a resistor. In order for the propagation constant to be real and the current along an antenna to be in-phase, the capacitances should not exceed the value determined by equality (2.42).

As an example, the V-dipole with an arm length $L = 1.5$ m and a radius 0.025 m was considered. Fifteen capacitances are included in each arm with spacing 0.1 m between loads (the first and last loads are placed at a distant of 0.05 m from the antenna end and the center). In order that the propagation constant remains real at frequencies up to 100 MHz, the capacitance closest to the generator is chosen to be 33.5 pF. The capacitances of other loads decrease along the antenna according to the linear law. As shown in Section 2.1, one can ensure the current distribution along an antenna close to linear and high level of matching with a cable.

Figure 3.12a shows the directivity of V-dipole with capacitive loads (curve 1) and without loads (curve 2) along the bisector of angular aperture with width 90°. For the sake of comparison, Fig. 3.12b shows similar curves for a linear dipole (curve 3 with loads, curve 4 without loads). The loads magnitudes and the antenna arm length are same. The calculations were executed in a frequency range from 100 to 500 MHz.

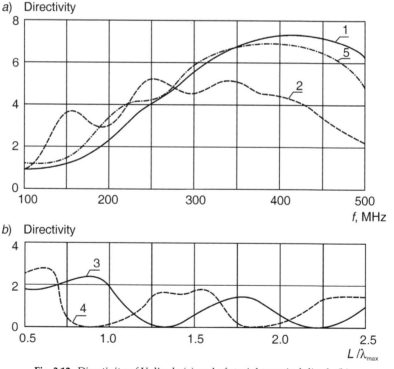

Fig. 3.12: Directivity of V-dipole (a) and of straight vertical dipole (b).

As seen from the figures, the directivity of a straight vertical dipole without loads in the direction perpendicular to the dipole axis quickly decreases at $L \approx (0.6–0.7)\lambda$. For a straight dipole with loads, this threshold value is found at $L \approx (1–1.2)\lambda$. The directivity of V-dipole along the angular aperture bisector is much higher and retains in a substantially wider frequencies range – from 350 ($L = 1.75\lambda$) to 500 MHz ($L = 2.5\lambda$). The loads increase the directivity of V-dipole by a factor between 1.4 and 2.8.

Figures 3.13 and 3.14 show the directional patterns of the earlier described V-dipole with loads (curve 1) and without loads (curve 2) in the antenna plane and the plane perpendicular to it. The calculated curves are compared with the experimental results (circles).

As was shown in Section 2.4, the mathematical programming method permits improvement in the electrical characteristics of an antenna with loads by optimal selection of the loads. Here, in order to calculate the zero approximation, it is expedient to use the method of a stepped impedance transmission line. In the considered example, the mathematical programming method allows increase in the directivity level of V-dipole with loads in the lower part of a frequency range (Fig. 3.12a, curve 5).

V-dipole with capacitive loads can be used as a directional antenna of VHF range.

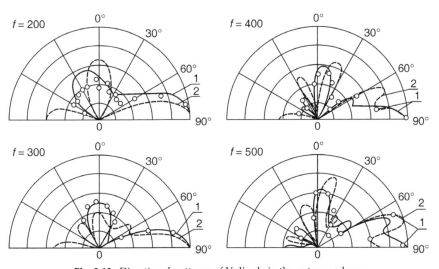

Fig. 3.13: Directional patterns of V-dipole in the antenna plane.

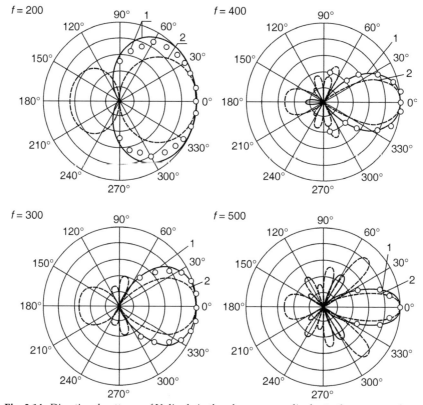

Fig. 3.14: Directional patterns of V-dipole in the plane, perpendicular to the antenna plane.

Directional Characteristics of Radiators with Capacitive Loads

4.1 Calculating the directional patterns of radiators with a given current distribution

In the previous chapters it was shown that antennas with in-phase current have higher electrical characteristics than ordinary antennas produced of metal wires with sinusoidal currents. In-phase current is a current whose phase is close to constant or at least does not change sign along the antenna wire. In this case, the amplitude of the current falls along the wire towards the free end of the antenna, in particular according to a linear or exponential law. To obtain in-phase currents, one can, for example, include concentrated capacitive loads along the antenna wire. Their capacitances should not exceed some values depending on the signal frequency.

The main attention before now has been paid so far to the problem of matching antenna impedance with the cable or the generator. As a result, the desired character of a directional pattern was obtained in a narrower frequency band than the required level of matching. The purpose of this chapter is to eliminate this drawback.

The possibility of antenna operation in a wide frequency range is determined by the properties of the radiator, whose length is larger than the wavelength. Therefore, the characteristics of antennas with a high length are of a great interest. Along with the level of antenna matching with the source of excitation, the most important electrical characteristic of any antenna is its directivity. The importance of creating the desired directional patterns follows in particular from the necessity to correct an error and abolish the drawback admitted in the synthesis of the radiators with concentrated capacitive loads.

The main characteristics of the directivity are two: first, this is the maximum directivity D_{max}, i.e., a ratio of maximum radiation intensity to its average value (to a value of isotropic radiation). In accordance with this formulation, the directivity is the coefficient of increasing total radiated power P, if a maximum power is radiated in all directions. In this case, a total power is

$$DP = 4\pi S_{max}, \tag{4.1}$$

where S_{max} is a power flux in the direction of maximum radiation, i.e., a total power equal to a product of a power in the direction of a maximum radiation by the area $S = 4\pi R^2$ of a sphere with a large radius R. The power flux per unit area is determined by the Poynting vector $S = [E, H]$, where E and H are intensities of electric and magnetic fields. In a cylindrical coordinate system, the radiated power of a typical antenna with an axial structure, symmetrical with respect to two planes, passing through coordinates origin, is

$$P = \int_0^{2\pi} d\varphi \int_{-\pi/2}^{\pi/2} S_{max} F^2(\theta, \varphi) d\theta = 4\pi \int_0^{\pi/2} S_{max} F^2(\theta) d\theta. \tag{4.2}$$

From (4.1) and (4.2) we find for an antenna directivity

$$1/D = P/(4\pi E_m^2) = \int_0^{\pi/2} \frac{S_{max} F^2(\theta)}{E_{max}^2} d\theta, \tag{4.3}$$

where E_{max} is the electric field in the direction of maximum radiation. In order to calculate the directivity magnitude of the antenna in an arbitrary direction, it is necessary to replace, in (4.3), E_{max} by a field E in this direction.

The second characteristic is the pattern factor (*PF*), which is equal to an average radiation level in a given range of angles (see Section 2.4, expression (2.56)). In order to increase the distance of radio communication, signals must be radiated along the earth's surface, for example, between angles θ from $60°$ to $90°$, and the share of power radiated in this angular sector can serve as the measure of an antenna quality. The pattern factor for the vertical plane is

$$PF = \frac{1}{K}\sum_{k=1}^{K} F(\theta_k), \tag{4.4}$$

where $F(\theta_k)$ is the magnitude of normalized vertical directional pattern at an angle θ in an angular sector from θ_1 to θ_K. If the directivity maximum of an antenna is large, but is located outside the necessary angular sector, then its magnitude is useless for increasing the radio communication range.

Expression (4.3) makes it possible to determine the antenna directivity, if the current distribution along the radiator is known. Calculation is based on relationships of power fluxes (of Poynting vectors) through each part of a sphere. Replacing the integration by summation, we obtain:

$$1/D = \sum_{m=1}^{N} \frac{F^2(\theta)}{E_{max}^2} \Delta \sin \theta_m. \tag{4.5}$$

Calculations are accomplished by means of (4.5). They are used known values of the current at points uniformly located along the antenna arm [20]. Let the axis of an antenna coincides with z-axis. The antenna center coincides with the origin of a cylindrical coordinate system. The arm is divided into N sections of equal length Δ. In the performed calculations, the value N was assumed to be 20, i.e., the value Δ in radians is equal to $\Delta = \pi/2: N = 0.0785$. The section fields are added together along the lines that are uniformly diverging in a vertical plane from the structure center. Power flows S_n in these directions pass through sections of a sphere of large radius with the same angular dimensions, i.e., the length of each section (between adjacent horizontal lines) is equal to Δ and the width of a section (between adjacent vertical lines) is equal to $\Delta \sin \theta_n$. The sphere of large radius is divided along azimuth into equal vertical strips, which create the same fields. The area of each section of this strip in the direction θ_n is equal to $\Delta^2 \sin \theta_n$ and the power flow through each section is $S_n = E_n^2 \sin \theta_n$. The total power flow is equal to $P = N \sum_{n=1}^{N} \Delta E_n^2 \sin \theta_n$. Thus the fraction of power flow in the horizontal direction is $E^2 = E_N^2$ accordingly. The directivity in this direction is equal to $D = NE_N^2/P = E_N^2 / \left(\sum_{n=1}^{N} \Delta E_n^2 \sin \theta_n \right)$, where N is the quantity of horizontal stripes on a sphere surface between angle $\theta = 0$ and $\pi/2$ (see Fig. 4.1), and n is a stripe number.

Variants of symmetrical metal radiators with sinusoidal current are given in Fig. 4.2: a—straight vertical dipole, b—V-dipole (θ_1 is the angle of an inclination of the antenna arm).

The electrical field of the straight dipole is determined by using the expression for the far field of Hertz dipole (elementary linear radiator) located along the axis z. The field in the far zone of the sinusoidal current $J(z) = J(0) \sin k (L - |z|)$ at the distance r is equal to

$$E_\theta = j \frac{60J(0)}{\sin kL} \cdot \frac{\cos(kL \cos \theta) - \cos kL}{\sin \theta} \cdot \frac{\exp(-jkr)}{r}. \tag{4.6}$$

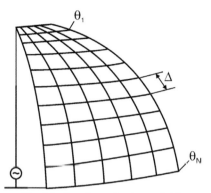

Fig. 4.1: Structure of antenna field.

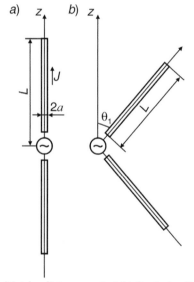

Fig. 4.2: Metal radiators: a—straight dipole, b—V-dipole.

Here $J(0)$ is the generator current, k is the propagation constant of the wave in air, L is the arm length. The second multiplier of the expression (4.6) is the directional pattern of the straight metallic dipole.

The electrical field of the upper arm of the metal V-dipole is equal to

$$E_\theta^{(1)} = AJ(0)\sin(\theta - \theta_1)\int_0^L \sin k(L - |z|)e^{jkL\cos(\theta - \theta_1)}dz = \frac{AJ(0)}{k^2\sin(\theta - \theta_1)}[e^{jkL\cos(\theta - \theta_1)} - \cos kL],$$

where $A = const\ (kL, \theta, \theta_1)$. Since the structure is symmetrical with respect to a plane passing through an origin of coordinates system and the current direction is the same, the field of the low arm is obtained by replacing signs

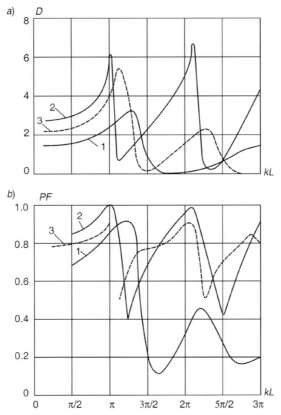

Fig. 4.3: Directivity (a) and pattern factor (b) of symmetric radiators with sinusoidal currents: of the straight vertical dipole (curves 1) and of V-dipole with the angle $\theta_1 = 30°$ and 45° (curves 2 and 3).

of θ and θ_1 by the opposite signs in the expression for $E_\theta^{(1)}$. Therefore, the total field of V-dipole is equal to

$$E_\theta = \frac{2AJ(0)}{k^2 \sin(\theta - \theta_1)}\{\cos[kL\cos(\theta - \theta_1)] - \cos kL\}. \qquad (4.7)$$

In Fig. 4.3 maximal directivity and pattern factor of the straight and V-dipoles are presented depending on kL. The sum of fluxes close to the earth helps in finding the field of the lower area of a sphere, i.e., to determine PF. When the length of the antenna arm is greater than 0.75λ, increasing number of sections with anti-phase current causes an additional maximum in the directional pattern, usually at great angles to the horizon. As a result, the signal along an earth is small and PF sharply decreases with increasing frequency, i.e., the appearance of additional maximum does not increase the communication range.

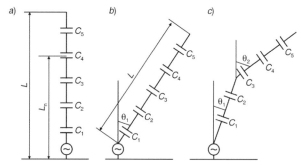

Fig. 4.4: Asymmetrical antennas with linear distribution of in-phase current: a—straight vertical antenna, b—V-dipole with straight arms, c—V-dipole with arms of two sections.

Connection of concentrated capacitive loads in the antenna wire allows creation of in-phase current with linear or exponential distribution and expansion of frequency range with a high level of matching. Variants of asymmetrical antennas with linear in-phase current are given in Fig. 4.4. In all cases, the capacitances of loads decrease towards the free antenna end in proportion to the distance from it. The electrical field of the upper arm of V-dipole with in-phase current in the first quadrant (in the angles range from $\theta = 0°$ to $\theta = 90°$) is equal to

$$E_\theta^{(1)} = AJ(0)\sin(\theta - \theta_1) \int_0^L (L - |z|)e^{jkLb}dz = \frac{AJ(0)\sin(\theta - \theta_1)}{k^2 b^2}[1 + jkLb - e^{jkLb}],$$

$$(4.8)$$

where $b = \cos(\theta - \theta_1)$. If the V-dipole with in-phase current is symmetrical, its total field is equal to

$$E_\theta = \frac{2AJ(0)\sin(\theta - \theta_1)}{k^2 b^2}(1 - \cos kLb).$$

$$(4.9)$$

In case of the straight vertical radiator, the magnitude b in this expression is $b = \cos \theta$. Equalities (4.8) and (4.9) generalize well-known expressions (4.6) and (4.7) in the case of in-phase currents.

The magnitudes D and PF of a straight dipole with in-phase current depending on an arm length are given in Fig. 4.5. As can be seen from this figure, they steadily increase with increasing arm length, but in the initial frequency interval (Fig. 4.6), V-antenna with in-phase current has a better directivity. This temporary advantage is caused by the higher directivity of V-antennas. We must, of course, bear in mind that these characteristics exist only if the in-phase nature of the current is kept.

If the arms of V-antenna are short (the arm length is less than 2π), the directivity increases when the angular aperture diminishes. When electrical lengths of the arms are longer, the directivity at first is more

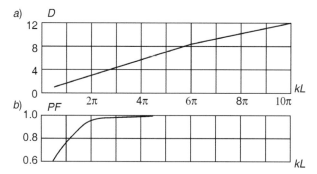

Fig. 4.5: Directivity (a) and pattern factor (b) of the straight vertical dipole with in-phase current.

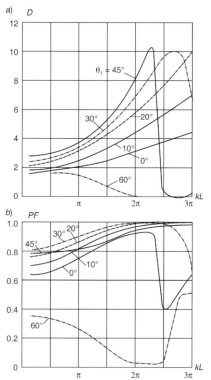

Fig. 4.6: Directivity (a) and pattern factor (b) of the V-dipoles with in-phase current and different angles of the arm inclination.

for an angular aperture at about 180° and then it weakly depends on the aperture. Comparison of directional properties of radiators with different current distribution shows that the radiators with in-phase currents allow obtaining the higher level of directivity and this level is preserved in a wide range of frequencies. If the arm of V-antenna is distinguished from the straight arm, for example, it consists of two sections, results depend on the angle of sections' inclination. When the sections with length $L_1 = L/2$ are deflected from the initial position in opposite directions under small identical angles (see Fig. 4.4c), the characteristics depend primarily on the section adjacent to the source (in Fig. 4.7, such curves are designated by two angles where the first angle corresponds to the section adjacent to the source). The expression for the field looks as

$$E = \frac{2AJ(0)}{k^2}\left\{ \frac{\sin(\theta - \theta_1)}{b_1^2}[1 - \cos kL_1 b_1 + kL_1 b_1] + \frac{\sin(\theta - \theta_2)}{b_2^2}[\cos kL_1 b_1(1 - \cos kL_1 b_2) + \sin kL_1 b_1(\sin kL_1 b_2 - kL_1 b_2)]. \right.$$

(4.10)

Here $b_i = \cos(\theta - \theta_i)$. This expression is easily generalized to N sections.

If the angular aperture between the sections is more than between the straight arms, the frequency range with the high level of directivity expands. If the arm of V-antenna consists of several sections, the result is ambiguous.

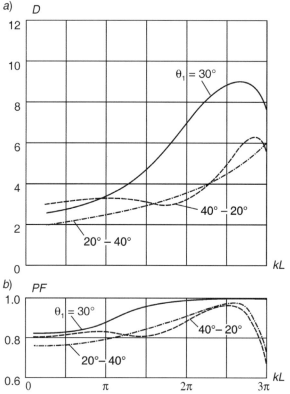

Fig. 4.7: Directivity (a) and pattern factor (b) of the V-dipole with the arm from one and two sections.

The obtained results allow to make a number of essential conclusions. Firstly, it is necessary to emphasize the usefulness of parameter *PF*. It detects frequencies at which the signal along the earth's surface is small (maximal directivity is great, but at a large angle to the horizon).

The directivity of V-dipoles, as a rule, is more than the directivity of straight radiators. If the current is sinusoidal, the directivity changes dramatically with frequency and other electrical characteristics of the antenna also change dramatically (*see* for example, Fig. 4.3, where the values of *D* and *PF* are given at the inclination angles 30° and 45°).

The radiators with in-phase currents allow to obtain a higher level of directivity, which is preserved in a wide range of frequencies. As can be seen from Figs. 4.5 and 4.6, straight vertical and V-dipoles maintain a smooth variation of directivity in a wide range with frequency ratio of at least 10. But in addition to that, the directivity of V-dipole is substantially more, especially for angles of inclination 30°–45° (for angular aperture 120°–90°).

The opportunity to use V-antennas, including complex structures, depends largely on their sizes and on the resistance variation near the series resonance. The arm length of such V-antennas at the point of series resonance can be reduced to 0.3–0.35 of the wavelength, i.e., the height of V-antenna with the aperture 90° may not exceed the height of the straight vertical dipole.

The radiation resistance in the general case is equal to

$$R_\Sigma = 20k^2h_e^2, \tag{4.11}$$

where h_e is effective length. If the vertical projection L of the arm of V-dipole is equal to the arm length of the straight dipole, then the effective length of the straight dipole is equal to $h_{e1} = \dfrac{2}{k} \tan(kL/2)$ and the effective length of in-phase V-dipole is $h_{e2} = L$. In the point of series resonance of the straight dipole $kL = \pi/2$, i.e., the effective length of this dipole is $h_{e1} = 2/k$ and the radiation resistance is $R_{\Sigma 1} = 80$ Ohm. The radiation resistance of V-dipole at this frequency is equal to $R_{\Sigma 2} = 20(\pi/2)^2 \cong 50$ Ohm. Equality of this resistance and he wave impedance of the standard cable is advantage of the in-phase V-dipole. Calculation shows that its radiation resistance is more stable under changing frequency. If the frequency changes from $0.95 f_0$ to $1.05 f_0$ where f_0 is the frequency of series resonance of the straight dipole, the value of $R_{\Sigma 1}$ increases from 68 to 94 Ohm and $R_{\Sigma 2}$ from 45 to 55 Ohm.

It is interesting to compare the characteristics of elementary antenna arrays consisting of two vertical V-antennas. The first array consists of antennas with in-phase current distribution, the second array consists of antennas with sinusoidal current distribution. In the first case, antennas are

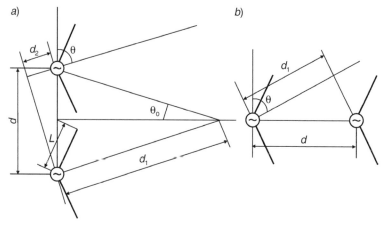

Fig. 4.8: Simplest arrays consisting of two vertical V-antennas, located one above other (a) and one behind other (b).

located one above the other (Fig. 4.8a) and in the second case, antennas are located one after the other (Fig. 4.8b). The arm length of each antenna is equal to $kL = 1.5\pi$, the angle of the arm inclination being $30°$, d is the distance between the antennas' centers. It is assumed that the fields created by each antenna have the same amplitude and phase and are equal to E_0. In the first case the antenna fields along the angle θ are equal to

$$E_1 = E_0 \exp(-jkd_1), E_2 = E_1 \exp(jkd_2),$$

$$E_1 + E_2 = E_1 \left[1 + \exp(jkd_2)\right] = 2E_1 \exp(jkd_2 / 2)\cos\frac{kd_2}{2},$$

where $d_2 = d \sin\theta_0$, $\sin\theta_0 = \cos\theta$, that is (see Fig. 4.8a)

$$E_1 + E_2 = 2 E_1 \exp(jkd_2/2) \cos\frac{kd\cos\theta}{2}.$$

Let's calculate D and PF of two arrays consisting of two radiators with in-phase current distribution and compare them with the values of D and PF for arrays with sinusoidal currents in the antennas. The directivity values of arrays are shown in Fig. 4.9a, the pattern factors—in Fig. 4.9b. In a first case the curves for the array with in-phase currents are denoted by 1, for the array with sinusoidal currents—by 2. In a second case the curves for the array with in-phase currents are denoted by 3, with sinusoidal currents—by 4.

Results of this comparison show additional interesting features of radiated structures with in-phase current. Directivity of these arrays is much higher than directivity of arrays with sinusoidal currents. The characteristics change smoothly with frequency.

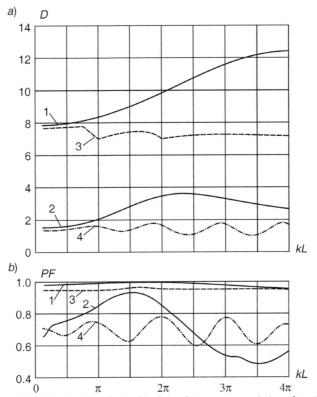

Fig. 4.9: Directivity (a) and pattern factor (b) of simplest arrays, consisting of two V-antennas located one above the other (curves 1 and 2) and one behind the other (curves 3 and 4).

4.2 Method of electrostatic analogy

The initial chapters of this book are devoted to improving electrical characteristics and expanding the frequency range of simple radiators consisting of one or two linear elements, in which are included or not included loads. Increasing number of linear elements in the antenna complicates the problem of its analysis and synthesis. These are in particular director, log-periodic and self-complementary antennas. Solving problems concerned with these antennas requires new approaches and new mathematical methods.

When solving mathematical problems, an analogous form of equations lends essential assistance. For example, the first two Maxwell's equations that became a basis for classical electromagnetic theory, allowed substantiating the principle of duality [21, 22]. In differential form, these equations look like these:

$$rot\vec{H} = \vec{j} + \varepsilon \frac{\partial \vec{E}}{\partial t}, rot\vec{E} = -\mu \frac{\partial \vec{H}}{\partial t}, \tag{4.12}$$

where $\vec{E}$ and $\vec{H}$ are strengths of electric and magnetic field respectively, $\vec{j}$ is a conduction current, ε and μ are permittivity and permeability of a surrounding medium. By replacing variables in accordance with the equalities $\vec{E}_1 = \sqrt{\varepsilon}\vec{E}, \vec{H}_1 = \sqrt{-\mu}\vec{H}, \vec{j}_1 = \sqrt{-\mu}\,j$ and introducing conventionally a magnetic conduction current, we obtain

$$rot\vec{H}_1 = \vec{j}_1 + \sqrt{-\mu\varepsilon}\frac{\partial \vec{E}_1}{\partial t}, rot\vec{E}_1 = \vec{j}_m + \sqrt{-\mu\varepsilon}\frac{\partial \vec{E}_1}{\partial t}, \tag{4.13}$$

where $\vec{E}_1$ and $\vec{H}_1$ are also strength vectors, and $\vec{j}_1$ is a conduction current, but in other units.

The obtained equations are completely symmetric with respect to $\vec{E}_1$, $\vec{H}_1$. An introduction of a magnetic conduction current makes theirs also symmetric with respect to electric and magnetic conduction currents, i.e., makes possible to treat (4.13) as equations not only for an electric radiator, but also for a magnetic radiator with a magnetic conduction current $\vec{j}_m$. Expressions for fields and characteristics of a magnetic radiator can be recorded using similar expressions for an electric radiator. For example, an input impedance of a magnetic radiator Z_m, whose shape and dimensions coincide with the shape and dimensions of an electric radiator, are equal to

$$Z_m = (120\pi)^2/Z_e. \tag{4.14}$$

Here Z_e is an input impedance of an electric radiator. A real embodiment of a magnetic radiator is a slot.

A similar method of solving problems is used in many other cases. For example, it is known an analogy between a picture of electrostatic field of charged conducting bodies located in a homogeneous and isotropic dielectric and a picture of constant currents in a homogeneous, weakly-conducting medium. In this case, bodies placed in the medium must have high conductivity and their shape and geometric dimensions must coincide with the shape and dimensions of the conducting bodies located in a dielectric.

If a picture of an electric field of linear charges is known, then, using the correspondence principle, one can construct a picture of a magnetic field of constant linear currents, provided the currents and charges are distributed in space identically. The difference between these images is only that lines of equal magnetic potential are located at places of lines of electric field strength, and lines of magnetic field strength are located at places of lines of equal electric potential [23].

Generalizing the principle of correspondence, it is advisable to compare electromagnetic fields created by high-frequency currents of linear radiators and electrostatic fields of charges located on linear conductors. Both fields are directly proportional to the magnitude of the current or the magnitude of the charge, and in the distant zone, they are inversely proportional to the distance from the source.

The offered method is based on an analogy between two structures consisting of high-frequency currents and constant charges. It is assumed that shapes and dimensions of radiators coincide with shapes and dimensions of conductors. In the case of several radiators a ratio of emf in their centers is equal to a ratio of charges placed on the conductors. The positive charge, equal to Q_0, is located on the conductor 0, which corresponds to the active radiator. The negative charges are located on the conductors i, corresponding to passive radiators. They are equal to $-Q_i$ and their sum is $\sum_{i=1}^{3}(-Q_i) = -Q_0$, i.e., the sum of all charges is zero and the conductors form an electrically neutral system. In this system

$$Q_i/Q_0 = C_{0i}/\sum_{(i)}C_{0i},\tag{4.15}$$

where C_{0i} is the partial capacitance between conductors 0 and i. As it follows from (4.15), the charges of the conductors i are directly proportional to the partial capacitances C_{0i} between these conductors and the conductor 0 (see, for example, [24]).

Equivalent replacement of a complex structure of high-frequency radiators by a structure with constant charges placed on conductors sharply simplifies the problem, reducing it to the electrostatic problem. In accordance with what has been said, it is natural to call the proposed method by the method of electrostatic analogy.

The considered method allows analysis of the problem in a common form, for example, to study and compare different laws of current distribution along the individual radiators. This is an undoubted advantage of the method. Characteristics of complex antennas are usually calculated using complex programs based on the method of moments. For the given problems, such a method is in essence a trial-and-error method. The method of moments does not permit comparison of antennas with different distribution of currents in a common form. Therefore, this method is not applicable here. The approximate method does not give exact results. But if this method is correct, i.e., corresponds to the physical essence of the problem, its accuracy is the same for different current distributions and the method allows a choice of the best option.

As stated in the Introduction, a reasonable sequence of solving each problem of synthesis requires that at the first stage, an approximate method of solving is developed, the results of which are used as initial values for the numerical solution of the problem by the method of mathematical programming. The method of electrostatic analogy can be used as an approximate method of this type.

It is expedient to consider the procedure of applying the electrostatic analogy method on a concrete example. As an example, the director-type antenna (Yagi-Uda antenna) described in [25] was adopted. It should be recalled that, as stated in the Introduction to the present book, the problem of optimizing the antenna characteristics by choosing dimensions by means of using mathematical programming methods was first solved with respect to the director antenna and this decision confirmed the correctness of the chosen approach. Work [25] was one of the first and most profound works devoted to optimization of the director antenna.

The antenna circuit is given in Fig. 4.10. The antenna consists of four metal radiators (active radiator, reflector and two directors). On Fig. 4.10, the antenna dimensions (in meters) are shown. They were determined by solving the optimization problem in a strict formulation. Let's start with the capacitance between the conductors. If the radii of the conductors i and 0 are the same and equal to $a = 0.001$ m and the lengths l_0 and l_i of these wires are slightly different from each other, then the partial capacitance C_{0i} between these wires in the first approximation is equal to

$$C_{0i} = \pi \varepsilon_0 l_i / \ln(b_i/a), \tag{4.16}$$

where ε_0 is the permittivity of a medium and b_i is the distance between the wires.

If to divide the wire 0 of the director antenna into three wires and to denote these wires by $0i$ indices, then the antenna circuit is divided into three circuits. Each circuit consists of two conductors: of wire i and wire $0i$

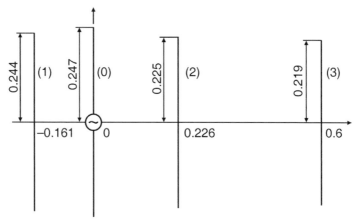

Fig. 4.10: Dimensions of the optimized director antenna from metal radiators.

(Fig. 4.11). The generator is also divided into three generators located in the centers of the wires $0i$. The values of their emfs are defined as

$$e_i = eQ_i/Q_0 \tag{4.17}$$

where e is the emf of the active radiator 0.

As is shown in the theory of folded antennas, the circuit of two parallel vertical wires located at a distance b_i can be divided into a dipole and a long line, open at both ends. The current at the center of each dipole is equal to

$$J_{id} = e_i / (4Z_{id}). \tag{4.18}$$

The reactive component of its input impedance is $X_{id} = -120 \ln(2L_0/a_{ei})\cot kL_0$, where $L_0 = l_0/2$ is the arm length of the active radiator, $a_{ei} = \sqrt{ab_i}$ is its equivalent radius. The current at the center of each wire of the long line is equal to

$$J_{il} = e_i / (2jX_{il}), \tag{4.19}$$

where $jX_{il} = -j120 \ln(b_i/a)\cot kL_i$ is the reactive input impedance of the long line with length $L_i = l_i/2$.

The current of the active and passive radiators is equal to the sum and difference of the dipole current J_0 and the current J_i of long line, where

$$J_0 = \sum_{i=1}^{3} (J_{id} + J_{il}), \quad J_i = J_{id} - J_{il}. \tag{4.20}$$

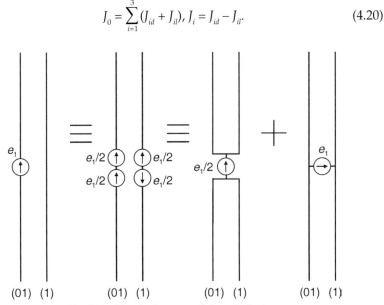

(01) (1)　　　　(01) (1)　　　(01) (1)　　　(01) (1)

Fig. 4.11: Circuit of two parallel vertical wires.

The amplitude and phase of the fields created by each radiator depend on its structure. The radiator structure determines the law of current distribution along it. If the radiator arm is a straight metallic conductor, its current is distributed according to the sinusoidal law $J(z) = J(0) \sin k (L - |z|)$. In this case, the distant field of the radiator is equal to

$$E = \frac{AJ(0)}{\sin \theta}[\cos(kL \cos \theta) - \cos kL], \tag{4.21}$$

where $A = j\dfrac{60}{\sin kL} \cdot \dfrac{\exp(-jkR)}{R}$, k is the propagation constant and R is the distance to the observation point. If the concentrated capacitive loads realize the in-phase current distributed along the radiator axis in accordance with a linear law $J(z) = J(0)(L - |z|)$, the far field of the radiator is equal to

$$E = \frac{AJ(0)\sin \theta}{\cos^2 \theta}[1 - \cos(kL \cos \theta)]. \tag{4.22}$$

From Fig. 4.10 it is obvious that the maximum radiation of the considered antenna is directed to the right, that is, towards the radiator 3. Since the radiator 1 is located on the left of the active radiator 0 at a distance b_1 from it, its field lags behind the field of the active radiator, first, per phase corresponding to a time of signal propagation from radiator 0 to the radiator 1 and secondly, per phase corresponding to the propagation time of the signal in the opposite direction from the radiator 1 to the radiator 0 (the signal of radiator 1 must come to the radiator 0 an angle θ, i.e., the path length between the wires is $b_1/\sin\theta$). The total phase difference is equal to $\psi_1 = - kb_1 (1 + \sin\theta)/\sin\theta$. Similarly, in the case of radiators 2 and 3, this phase difference is equal to $\psi_2 = - kb_2 (\sin\theta - 1)/\sin\theta$ and $\psi_3 = - kb_3 (\sin\theta - 1)/\sin\theta$, respectively.

The described procedure permits calculation of the total field of the director antenna shown in Fig. 4.10. In accordance with (4.17), emfs of the different radiators are equal to $e_1 = 0.388\ e$, $e_2 = 0.335\ e$, $e_3 = 0.277\ e$. The total field of this antenna with radiators in the form of straight metal wires is

$$E = \frac{AJ(0)}{\sin \theta}\sum_{i=1}^{3} e_i \left\{ \left(\frac{1}{4Z_{id}} + \frac{1}{2Z_{il}} \right)[\cos(kL_0 \cos\theta] - \cos kL_0] + \left(\frac{1}{4Z_{id}} + \frac{1}{2Z_{il}} \right)\exp(j\psi_i)[\cos(kL_i \cos\theta) - \cos kL_i] \right\}. \tag{4.23}$$

The emfs of the radiators do not change.

The inclusion of concentrated capacitive loads along the linear radiators, the magnitudes of which vary in accordance with the linear or exponential law, permits to create radiators with in-phase current. While retaining the dimensions of the radiators and the distances between them, we get the director antenna as shown in Fig. 4.12. The total field of such an antenna is calculated by the formula

$$E = \frac{AJ(0)}{\sin\theta}\sum_{i=1}^{3} e_i \left\{ \left(\frac{1}{4Z_{id}} + \frac{1}{2Z_{il}} \right)\left[1 - \cos(kL_0 \cos\theta)\right] + \left(\frac{1}{4Z_{id}} - \frac{1}{2Z_{il}} \right)\exp(j\psi_i)\left[1 - \cos(kL_i \cos\theta)\right] \right\}.$$

(4.24)

The results of calculating directivity and pattern factor of the director antenna with straight metal wires are shown in Fig. 4.13 (curve 1). Since

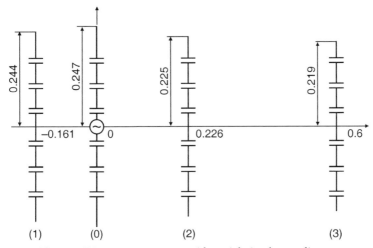

Fig. 4.12: Director-type antenna with straight in-phase radiators.

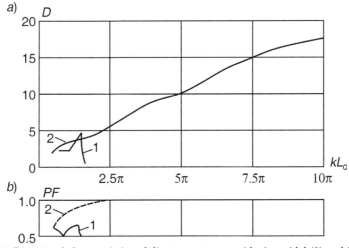

Fig. 4.13: Directional characteristics of director antennas with sinusoidal (1) and in-phase (2) currents.

here an approximate calculation procedure is used, these results are not identical to the results presented in [25], but are similar to them. Practically the antenna operates at the same frequency. The results of calculating these characteristics for the antenna with in-phase currents are shown in Fig. 4.13 (curve 2). They speak for themselves. This antenna operates over a wide frequency range and its directivity steadily and smoothly increases with increasing frequency, that is, the quality factor of this antenna is small. Of course, we must bear in mind that these characteristics are valid only if the current is in-phase. But the frequency ratio of antennas with capacitive loads with a high level of matching is a value to the order of 10.

As has already been said, the method of electrostatic analogy is based on the resemblance of a structure consisting of high-frequency currents and a structure consisting of constant charges. Comparison of electromagnetic fields created by high-frequency alternating currents of linear radiators with electrostatic fields of charges placed on linear conductors of electrically neutral system shows similarity of the mathematical structures of both fields.

This method allows to propose a simple and effective procedure for calculating the directional characteristics of the director antennas from linear radiators. The procedure uses knowledge about the basic antenna dimensions and current distributions along the radiators. It does not require detailed information on the types and magnitudes of concentrated loads. As calculations were shown, the director-type antennas from the linear radiators with in-phase currents provide a high directivity and smooth variation of characteristics in a wide frequency range. The results obtained with help of this method can be used to solve the problem of optimizing various director antennas by mathematical programming methods.

Similar results can be obtained for V-radiators. In this case, the operating range decreases, but *TWR* increases at concrete frequencies. The circuit of such director-type antenna with V-radiators is given in Fig. 4.14. Calculations are performed for antennas with the same lengths of radiators and the same distances between them as in the antenna shown in Fig. 4.11. The angles between the arms of the radiators and the verticals in each antenna are the same and are equal to $10°$, $20°$, and $30°$, respectively. In this case, the total antenna field is equal to

$$E = \frac{AJ(0)}{\sin\theta}\sum_{i=1}^{3}e_i\left\{\left(\frac{1}{4Z_{id}}+\frac{1}{2Z_{il}}\right)\left[1-\cos(kL_0\cos\theta)\right]+\left(\frac{1}{4Z_{id}}-\frac{1}{2Z_{il}}\right)\exp(j\psi_i)\left[1-\cos(kL_i\cos\theta)\right]\right\}.$$

$$(4.25)$$

The directional characteristics of the director antennas with V-radiators are shown in Fig. 4.13. The value of *PF* does not drop below 0.85. The results show that these antennas, in comparison with the antennas of

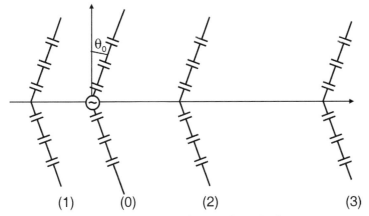

Fig. 4.14: Director antenna from in-phase V-radiators.

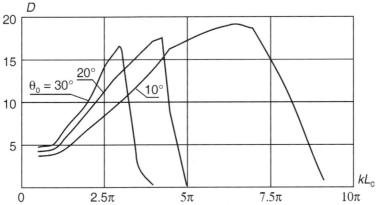

Fig. 4.15: Directivity of director antennas from in-phase V-radiators.

straight in-phase radiators allow increasing directivity but with a narrower frequency range. In this case the directivity and frequency range depend on the inclination of the antenna arm: if this angle increases, the directivity at low frequencies also increases and the frequency range becomes narrower.

As indicated in the Introduction, a reasonable sequence of solving each problem recommends at the first stage to propose and develop an approximate method of analysis and then use its result as initial values for the numerical solution of the problem by methods of mathematical programming. The method of electrostatic analogy allows to take the first step. The results obtained with its help can be used to solve the problem of optimizing various director antennas by mathematical programming methods.

4.3 Decreasing dimensions of log-periodic antennas

As shown in the previous section, the in-phase current in the radiators of director antenna provides a higher directivity over a wide frequency range. Such radiators can also be used as elements of a log-periodic antenna. As an approximate method of analysis, one can apply the method of electrostatic analogy between two structures consisting of currents and charges in the same system of wires. We will assume that the ratio of emfs in the radiators centers is equal to the ratio of charges placed on these conductors. The sum of all the charges is zero and the conductors form an electrically neutral system. In this system the relation (4.15) is valid. From this relation it follows that the negative charge $- Q_i$ of the passive conductor i is directly proportional to the partial capacitance C_{0i} between this conductor and the active conductor 0. The capacitance between two parallel conductors of radius a is found from (4.16).

In accordance with the well-known method of analyzing the log-periodic antenna [26], we consider the active region of this antenna consisting of three radiators (Fig. 4.16) and determine the fields of these radiators when the emf e is located in the center of the radiator 0 (of middle radiator). In accordance with Kirchhoff's law, other generators must be short-circuited. These are emfs in the centers of other radiators and at the ends of sections of distribution long line near the excitation points of neighboring radiators. Let the arm length of the middle radiator be equal

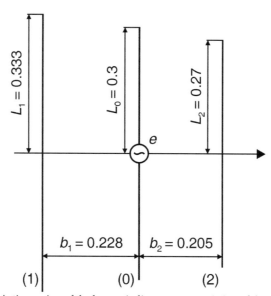

Fig. 4.16: Active region of the log-periodic antenna consisting of three radiators.

to $L_0 = \lambda/4$, the arm length of the left (longer) radiator is equal to $L_1 = \lambda/(4\tau)$, and the arm length of the right (shorter) radiator is $L_2 = \lambda\tau/4$, where τ is the denominator of the geometric progression, according to which the dimensions of the radiators change. Another parameter σ of the antenna is $\sigma = 0.25(1-\tau)\cot \alpha$, where α is the angle between an antenna axis and a line passing through the ends of the radiators. The magnitude σ is the distance in wavelengths between a half-wave radiator and a smaller neighboring radiator: $b_2 = \sigma\lambda$. Correspondingly the distance between a half-wave radiator and a longer neighboring radiator is $b_1 = \sigma\lambda/\tau$. As shown in [27], a generalization of the information available in the literature leads to the conclusion that change of electrical characteristics of a log-periodic antenna with metal dipoles within the frequency range from f to $f\tau$ is minimum when $\sigma/\tau = 0.19$. If, to consider that this relation is true for the antenna under study and for definiteness to assume that $\tau = 0.9$, then $L_1 = L_0/0.9 = 0.278\lambda$, $L_2 = L_0 \cdot 0.9 = 0.225\lambda$, $\alpha = 0.146$, $b_1 = 0.19\lambda$, $b_2 = 0.171\lambda$.

The procedure of analyzing the active region of a log-periodic antenna is similar to the procedure of analyzing the director antenna. For greater simplicity and clarity, it is necessary to repeat some details. To apply later on the theory of folded antennas, the active radiator (wire 0) is divided into two parallel wires (with numbers 01 and 02) and then the obtained circuit is divided into two circuits (wires with numbers 01 and 1 in the first circuit, wires with numbers 02 and 2 in the second). In accordance with the method of electrostatic analogy of two structures, i.e., in accordance with the physical content of the problem, it is assumed that the ratio of emfs at the centers of the wires 01 and 02 is equal to the ratio of partial capacitances C_{01} and C_{02}, i.e., $e_1 = 0.52e$, $e_2 = 0.48e$.

As shown in the theory of folded antennas, the structure of two wires located at a close distance b_i can be divided into a dipole and a long line open at both ends. The current at the center of each dipole is equal to $J_{id} = e_i/(4Z_{id})$. The current at the center of each conductor of the long line is $J_{il} = e_i/(2Z_{il})$. Accordingly, the current in the center of the active radiator is the sum of the dipole current with the current $J_0 = \sum_{i=1}^{2}(J_{id} + J_{il})$ of long line,

and the current in the center of the passive radiator is the difference between these magnitudes, $J_i = J_{id} - J_{il}$.

In the case of metal wires in the first approximation, the input impedance of the dipole is $Z_{id} = R_d + jX_{id}$, where $R_d = 80 \tan^2 (kL_0/2)$, $X_{id} = -120 \ln (2L_0/a_{ei}) \cot kL_0$, $a_{ei} = \sqrt{ab_i}$ is the equivalent radius of the dipole. The input impedance of the line is $jX_{il} = -j\,120 \ln(b_i/a)\cot kL_i$. If each element of the log-periodic antenna is a radiator with concentrated capacitive loads, whose magnitudes change along the radiator axis in accordance with the linear law, i.e., the current along each radiator is in-phase current, then the active and reactive components of the dipole input impedance are equal to $R_d = 20\, k^2\, L_0^2$ and $X_{id} = -120\, \ln(2L_0/a_{ei})/kL_0$. The input impedance of the long line is $jX_{il} = -j120\, \ln(b_i/a)/kL_i$.

The field amplitude and phase of each radiator depend on its structure and location in the antenna. If the radiator arm is a straight metal wire, the current along it is distributed according to the sinusoidal law. In this case its field in the far zone is calculated in accordance with (4.21). If the concentrated capacitive loads located along the radiator axis allow realization of in-phase current with a linear amplitude distribution, the field in the distant zone of the radiator is calculated in accordance with (4.22).

If the structure consists of several radiators close in size, the analysis of its characteristics confirms the possibility of creating a simple structure of a log-periodic directional antenna. Let's consider the distant field of a director antenna, as shown in Fig. 4.16, which is equivalent to the active region of the log-periodic antenna. The arm length of the middle (active) radiator is 0.3 m, i.e., the wavelength at the frequency of first series resonance is 1.2 m. The radii a of all conductors are same: $a = 0.001$ m. The value τ is equal to 0.9. It is obvious that the maximum radiation of the antenna should be directed to the right (to the radiator 2). Since the radiator 1 is located to the left of the active radiator 0, at a distance b_1 from it, the signal of radiator 1 is lagging from the signal of radiator 0, since pass first the distance from the active radiator to the passive radiator 1, and secondly in the opposite direction, from the radiator 1 to the radiator 0 (the signal of radiator 1 must come to the radiator 0 at the angle θ, i.e., the path length between the wires is $b_1/\sin\theta$). The total phase difference is equal to $\psi_1 = -kb_1\,(1 + \sin\theta)/\sin\theta$. Similarly, in the case of radiator 2 this phase difference is equal to $\psi_2 = -kb_2\,(\sin\theta - 1)/\sin\theta$.

The total field of the described antenna structure consisting of radiators with in-phase currents depending on angle θ is equal to

$$E = \frac{AJ(0)\sin\theta}{\cos^2\theta} \sum_{i=1}^{2} e_i \left\{ \left(\frac{1}{4Z_{id}} + \frac{1}{2Z_{il}} \right) \left[1 - \cos\left(kL_0 \cos\langle\theta - \theta_0\rangle \right) \right] + \right. $$

$$\left. \left(\frac{1}{4Z_{id}} - \frac{1}{2Z_{il}} \right) \exp(j\psi_i) \left[1 - \cos\left(kL_i \cos\theta \right) \right] \right\}. \tag{4.26}$$

Directivity is determined by the expression

$$D = |E(\pi/2)|^2 / \sum_{n=1}^{N} [\,|E(\theta_n)|^2\, \Delta \sin\theta_n\,], \tag{4.27}$$

where Δ is an interval between neighboring values θ_n, N is a number of intervals between $\theta = 0$ and $\theta = \pi/2$.

The results of calculating the directivity of the structure shown in Fig. 4.16 are shown in Fig. 4.17, depending on the electrical length kL_0 of the active radiator arm. Curve 1 shows the directivity of a structure with in-phase currents in each element. Curve 2 shows, for comparison, the directivity of the structure with sinusoidal currents. Radiators with concentrated capacitances distributed in accordance with a linear or exponential law along each arm allow a high level of matching with a cable in a range with a frequency ratio of about 10. Therefore, the graph demonstrates the structure directivity in the range from $kL_0 = \pi$ up to $kL_0 = 10\pi$.

Active regions of a log-periodic antenna slightly different in size with an arm length of active radiator, equal to 0.27 m and 0.333 m respectively, are shown in Figs. 4.18a and 4.18b. Their directivities are also shown in Fig. 4.17: curve 3 corresponds to Fig. 4.18a, curve 4—to Fig. 4.18b. The directivity magnitudes in Fig. 4.17 for specific values kL correspond to the same frequencies for all three schemes. From the Fig. 4.17 it follows that the directivity values of different schemes at identical frequencies

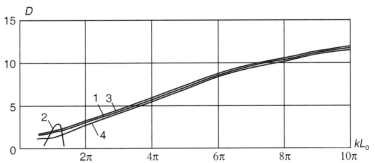

Fig. 4.17: Directivity of director antenna from three radiators.

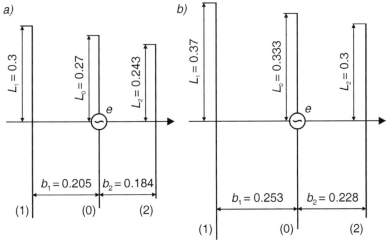

Fig. 4.18: Active regions of smaller (a) and larger (b) neighboring radiators.

are close to each other. This is natural, since the directivity of each active region slowly increases with increasing frequency. A small increase of the active radiator length leads, at the same frequency, to a small increase in the resonant wavelength, i.e., to a slight decrease of the resonance frequency and directivity.

The obtained results show that in the log-periodic antenna structure, each active region, consisting of one active and two passive radiators with in-phase currents, provides a high directivity in a wide frequency range. Neighboring radiators have a similar magnitude of directivity. The radiation direction is the same. The power propagates along the distribution line from short to long elements (in Figs. 4.16 and 4.18 from right to left) while the radiated signal propagates in the opposite direction. The total path difference is large enough. For example, for signals radiated by a half-wave radiator and an adjacent shorter radiator, this difference is equal to 0.38λ that is close to half wavelength. The crossing of wires in the distribution line in the interval between the radiators is making it possible to sharply reduce this difference.

A known log-periodic antenna with a sinusoidal current distribution along the radiators has the property of automatic cut-off of currents, i.e., each antenna part (active region) radiates a signal in a narrow frequency band. Outside this band and outside the active region, the signal quickly decreases. A wide frequency range is provided by a large antenna length, which is equal to the sum of the lengths of the active regions. Attempts to reduce this antenna length by disturbing a geometric progression and increasing the number of radiators lead to a small decrease in length and a sharp deterioration of electrical characteristics. More useful methods are, firstly, two-fold use of each active region by means of linear-spiral radiators, and secondly, the use of an asymmetric log-periodic antenna with a coaxial distribution line [27]. These methods can reduce the antenna length by 25–30 percents. This antenna is described in Chapter 11.

The comparative analysis performed by means of a simple and effective procedure, based on the method of electrostatic analogy, shows that the replacement of metal radiators by radiators with concentrated capacitive loads allows to provide a higher directivity in a wide range with a frequency ratio k_f of the order 10, that is greater than a frequency ratio of present-day log-periodic antennas. The calculation shows that for identical maximum wavelengths, the maximal length of radiators with loads should be longer approximately one-and-a-half times than the maximal length of conventional radiators. But the length of new antenna for the same frequency range may be equal to the length of a maximal radiator, i.e., substantially shorter, which is much more important than the transverse dimensions of antenna.

5

Adjustment of Characteristics of Self-complementary Antennas

5.1 Volume self-complementary radiators

As shown in previous chapters, use of concentrated capacitive loads in an antenna wire makes possible creation in it of the in-phase current and significant expansion of a frequency range, in which high electrical characteristics are provided. Similar problems can be solved also by means of complementary and self-complementary antennas.

The complementarity principle comes from to an interconnection between scattering properties of the metal and slot radiators of the same shape and dimensions. Two infinite plates (each plate is an aggregate of the metal and slot antenna), in which metal antenna is replaced by slot antenna and vice versa are named by complementary structures [28, 29].

Influence of the complementary principle is connected with the duality principle, based on symmetry of Maxwell's equations with respect to the quantities $\sqrt{\varepsilon_0}\,\vec{E}$ and $\sqrt{-\mu_0}\,\vec{H}$. The existence of metal (electric) and slot (magnetic) antennas follows from this symmetry. These antennas create fields of identical form as a result of a mutual replacement of appointed quantities. If one can place one metal antenna and one slot antenna on a flat metal sheet of infinite dimensions (it is assumed that the sheet is infinitely thin and has infinite conductivity) and these antennas occupy the entire sheet, the antennas are called complementary. As was shown, input impedances of the complementary antennas are related by the expression

$$Z_s = (60\pi)^2 / Z_e.$$ (5.1)

Here magnitude Z_s is input impedance of the slot radiator, equal to input impedance of the adjacent metal radiator, which occupies the rest of the infinite sheet. Magnitude Z_e is input impedance of the metal radiator, whose shape and dimensions coincide with shape and dimensions of the slot. If these antennas are located on the infinite sheet and have the same shape and dimensions, they are called self-complementary and

$$Z_s = Z_e = 60\pi, \tag{5.2}$$

i.e., the input impedance of each radiator is independent of frequency, purely active and equal to a wave impedance 60π.

In 1957 Rumsey formulated the constructive feature of these antennas as follows: if the shape of the antenna were such that it could be specified entirely by angles, its performances do not depend on frequency [30]. Special properties of such structures make them unique. These properties are partially preserved, if dimensions of a metal sheet are finite.

The first considered structure of such an antenna is presented in Fig. 5.1a. It consists of metal and slot symmetrical radiators of the same shape and dimensions. This antenna differs from self-complementary by finite dimensions of the structure. If her triangular metal surfaces to replace by wires dispersing out of the feed point (out of the lower vertex of the triangle), the antenna will take a form shown in Fig. 5.1b. Her characteristics are close to characteristics of the antenna shown in Fig. 5.1a.

Antenna consisting of a metal and a slot radiator of the same shape and dimensions should not necessarily be flat. For example, antennas placed on the surface of a circular cone, which is shown in Fig. 5.2, satisfy this formulation. As indicated in [31], if the angular width of the metal strip and the slot in the case of a conical two-thread helix is the same, then we can talk about the self-complementary structure, implying the identity of the cone surface covered with a metal shell and free from it.

In the general case, an expression similar to (5.1), is valid for the input impedance of a symmetrical bilateral slot antenna of arbitrary shape and dimensions, located on a circular metal cone of infinite length and excited in its vertex (Fig. 5.3). In order to derive this expression, one can use the duality principle. Consider a symmetrical magnetic V-radiator (Fig. 5.4a) located in free space. Its radiation resistance R_M is related with the radiation resistance R_e of similar shape and dimensions electric radiator by the expression

$$R_M = (120\pi)^2 / R_e. \tag{5.3}$$

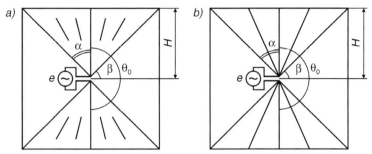

Fig. 5.1: Asymmetric (a) and symmetric (b) self-complementary radiators.

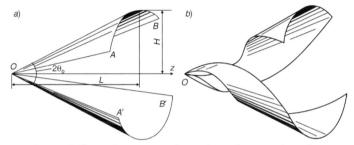

Fig. 5.2: Self-complementary radiators located on a conic surface.

Fig. 5.3: Slot on a surface of a circular cone.

This expression is well-known and follows from a comparison of the radiated powers. If to compare with each other the oscillating powers of these radiators, then similarly (5.3)

$$Z_M = (120\pi)^2 / Z_e, \tag{5.4}$$

where Z_M and Z_e are input impedances of the magnetic and electric radiators respectively.

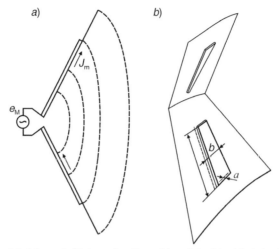

Fig. 5.4: Magnetic V-shaped radiator (a) and double-sided slot (b).

To go from the magnetic V-radiator to the slot antenna, we will divide each arm of the magnetic V-radiator by means of a conic metal surface of a circular cross-section, which passes through its axis. As shown in [32], the field strength surfaces, along which the force lines are located, are circular cones. The cones axes coincide with the bisector of the angle between the V-radiator arms. Since the shape of magnetic force lines coincides with the surface shape, the radiator field does not change as a result of introducing the metal surface.

The new metal surface divides V-radiator into two radiators located at its different sides (inside and outside the conical surface). Since magneto-motive force e_M exciting the radiator and the oscillating power $P = e_M J_M$ created by the generator do not change, the fractions of magnetic current J_M in each of the new radiators are equal to the fractions of power radiated into a corresponding part of the space. For definiteness, we assume that m is the fraction of power in a smaller part of the space (within the solid angle) and $(1-m)$—in the greater.

The input admittance of the magnetic radiator located inside the cone is

$$Y_1 = e_M/(mJ_M) = Y/m, \tag{5.5}$$

where $Y = 1/Z_M$ is the total admittance of the original radiator. Let the cross-section of the internal radiator be in the shape of a slightly curved parallelepiped with sides L, b and a, and $b>>a$ (Fig. 5.4b). The width b is parallel to the metal surface and has the shape of an arc. This radiator is equivalent to a one-sided slot of width b. The width b varies along the slot surface.

The second radiator is equivalent to a similar one-sided slot with the current $(1 - m)J_M$. Its input admittance is

$$Y_2 = e_M/[(1 - m)J_M] = Y/(1 - m).\qquad(5.6)$$

If we replace both slots by a single bilateral slot and assume that its admittance is equal to the sum of the both slot admittances, then

$$Y_s = Y_1 + Y_2 = Y/[m(1 - m)].\qquad(5.7)$$

The input impedance of the bilateral slot is equal to $Z_s = 1/Y_s$, or taking into account (5.4) and (5.7)

$$Z_s = m(1 - m)Z_M = (120\pi)^2\, m(1 - m)/Z_e.\qquad(5.8)$$

In particular, if the assumed magnetic radiator is straight, the conical surface turns into a vertical plane, which divides the space into two equal parts. Accordingly, the current fraction in each of the anew formed radiators is equal to $m = 1/2$, and equality (5.8) turns into the expression (5.1). If the metal and slot radiators are identical in shape and dimensions, then

$$Z_s = Z_e = 120\pi\,\sqrt{m(1 - m)},\qquad(5.9)$$

i.e., volume self-complementary structures have properties similar to those of flat structures, particularly constant and purely resistive input impedance, which is independent on frequency.

Performed analysis shows that the choice of the surface for the placement of the self-complementary radiated structures is not random. That is a surface of revolution, particularly a circular cone, which in the limit turns into a plane. This result was for the first time presented in [32]. The shape of the metal and slot radiators located on a circular cone, can be different—similarly to the shape of radiators located on the plane. In the simplest shape, the radiator boundary coincides with the cone generatrix (see Fig. 5.2a). Question of the coefficient m in the general case stays open.

5.2 Self-complementary radiators on a conic surface

The metal radiator placed on a circular cone may be located at an acute angle θ_0 to the ground. This allows to increase dimensions (the arm length) and directivity of the antenna, using supports of the same height. In order for the dimensions of the metal and the slot radiators to coincide, the distance between supports should be twice of their height. Then the points A and B and their mirror images coincide with the vertices of a square (see Fig. 5.2a). The conical problem exemplifies the three-dimensional problem of calculating the electric field of charged bodies. Such problems are difficult.

They are get simplified if all the quantities characterizing the field depend only on two coordinates. In this case, the field is called a plane-parallel. It is of interest to use the results of solving a two-dimensional problem for the calculation of the field in a three-dimensional problem, when the metal bodies' position resembles a two-dimensional version. In this case the surfaces of the conductive bodies coincide with the surfaces of equal potential and the electrostatic problem solution in accordance with the uniqueness theorem must satisfy Laplace's equation.

Conical and cylindrical problems are compared with each other in [33], where it is shown that the Laplace's equation stays valid in transition from one problem to another, if the replaced variables are related by equations:

$$\rho = \tan(\theta/2), \ \varphi_c = \varphi. \tag{5.10}$$

Here ρ and φ_c are cylindrical coordinates, θ and φ are spherical coordinates. The cylindrical problem is the problem about the circular metal cylinder with longitudinal slots of constant width, which cut through in it. The conical problem is the problem of two converging shells. The capacitance per unit length and the wave impedance of the cylindrical line (see, e.g., [24] or [34]) are equal to

$$C = \varepsilon K \left(\sqrt{1-k^2} \right) / K(k), \ W = 120\pi \ K(k) / K \left(\sqrt{1-k^2} \right), \tag{5.11}$$

where $K(k)$ is the complete elliptic integral of the first kind with argument $k = \tan^2(\beta/2)$, and β is the angular half-width of the slot. Thus, C and W depend only on the angular width of the slot 2β and, accordingly on the angular width of the metal shell $2\alpha = \pi - 2\beta$. Since they do not depend on the cylinder radius, both expressions are valid for conical shells forming the conical line.

Magnitudes of C and W are constant along the conical line, i.e., the long line is uniform. It is known that when the line length increases, the input impedance of such a line tends towards its wave impedance Z_1. Input impedance of the line located at the cone vertex tends towards the magnitude

$$Z_1(k) = 120\pi \ K(k)/K \left(\sqrt{1-k^2} \right). \tag{5.12}$$

Described conic structure is on the one hand a line and on the other, it is a metal V-radiator, whose arms perform in the form of converging shells located along the surface of a circular cone. Finally, this same structure can be regarded as a symmetrical slot V-antenna cut in the conic screen. If $Z_e(2\alpha)$ is the input impedance of the metal V-radiator with the arm angular width 2α, and $Z_s(2\beta)$ is the input impedance of the slot V-antenna with the arm angular width 2β, then it follows that

$$Z_e(2\alpha) = Z_s(2\beta) = Z_l(k) = 120\pi \, K(k)/K\left(\sqrt{1-k^2}\right). \tag{5.13}$$

If the angular widths of the metal shell and the slot are the same, i.e., $2\alpha = 2\beta = \pi/2$, then $k = \tan^2(\pi/8) = 0.172$, $K(k)/K\left(\sqrt{1-k^2}\right) = 0.5$, and consequently

$$Z_e(\pi/2) = Z_s(\pi/2) = Z_l(0.172) = 60\pi. \tag{5.14}$$

Let the metal shell and the slot have different widths. It is necessary to keep in mind that the input impedance of the slot is the input impedance of the shell located next to the slot. If the shell width is $2\alpha = 2\pi/3$, and the slot width is $2\beta = \pi/3$, then $k = \tan^2(\pi/12) = 0.0718$, $K(k)/K\left(\sqrt{1-k^2}\right) = 0.391$, i.e., $Z_s(2\pi/3) = Z_s(\pi/3) = 120\pi \cdot 0.391$. If the slot and shell widths are equal to $2\beta = 2\pi/3$ and $2\alpha = \pi/3$, then $k = \tan^2(\pi/6) = 0.333$, $K(k)/K\left(\sqrt{1-k^2}\right) = 0.641$, i.e., $Z_e(\pi/3) = Z_s(2\pi/3) = 120\pi \cdot 0.641$. Therefore,

$$Z_e(2\pi/3) \cdot Z_s(2\pi/3) = (120\pi)^2 \cdot 0.391 \cdot 0.641 = (60\pi)^2. \tag{5.15}$$

Expression (5.15) relates the slot impedance and the impedance of the metal radiator that is identical to slot in shape and dimensions.

It is easy to verify that this expression satisfied radiators of any width. Comparing (5.15) and (5.14) with (5.8) and (5.9) shows that the value m in expressions (5.8) and (5.9) is equal to $1/2$, i.e., the same power is radiated into the inner and outer solid angles, regardless of the opening angle of the cone and the angular width of the slots. From this, in particular, it follows that the conic radiator produces a directional radiation and its directivity is maximum along the axis of its aperture.

Two versions of the slots cut in a metal cone are shown in Fig. 5.2. In the first one, the slot boundary coincides with the generatrix of the cone. In the second one, the slot boundary is a helix line and the cone with the slots forms a two-thread helix, excited at the vertex. Strictly speaking, formulas (5.13)–(5.15) are valid only for the first version of the slot. However, if the widths of the shells and the slots are the same in both versions, one can be convinced that the capacitances per unit length and the wave impedances, and therefore the input impedances of the radiators, are the same for both versions, at least for infinitely long symmetric slots on a circular cone.

5.3 Self-complementary radiators on a parabolic surface

The magnetic radiator, which was used in derivation of (5.8) coincides with a surface of magnetic field strength. It can be not straight but a curved one. For example, it can have the shape of a parabola. Then a metallic surface, which coincides with the surface of the magnetic field strength, will have a shape of a circular paraboloid with a vertex at the point of excitation (Fig. 5.5a). In the general case, that is a surface of revolution with a curved generatrix. On such a surface as well as on a circular cone, which is its particular case, also it is possible to place the self-complementary radiators, to which expression (5.9) is applicable. If radiators are not self-complementary, equality (5.8) is valid.

Parabolic and cylindrical problems are compared with each other in [35], where it is shown that the Laplace's equation in transition from one problem to another stays valid, if the replaced variables are related by equalities:

$$\rho = \sigma, \ \varphi_c = \psi. \tag{5.16}$$

Here ρ and φ_c are the cylindrical coordinates, σ and ψ are the parabolic coordinates. Hence, the Laplace's equation holds true in transition from the parabolic problem to the cylindrical one, if expressions (5.16) are valid.

The analysis of such a structure is facilitated by the use of the parabolic coordinate system (σ, τ, ψ). This is a system of orthogonal curvilinear coordinates [36]. Their coordinate surfaces are confocal paraboloids of rotation $(\sigma = const, \tau = const)$ with focuses at the coordinates' origin and half-planes $(\psi = const)$ passing through the axis of rotation (see Fig. 5.5b). Rectangular coordinates are related to parabolic coordinates by the expressions:

$$x = \sigma \tau \cos \psi, \ \ y = \sigma \tau \sin \psi, z = (\tau^2 - \sigma^2)/2. \tag{5.17}$$

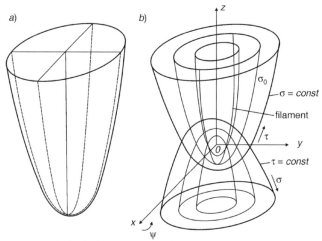

Fig. 5.5: Paraboloid of rotation (a), and parabolic coordinates system (b).

In this case the wires are located along the surface of a paraboloid, or more exactly along the curves of intersection of this surface by half-planes passing through the axis of rotation ($\sigma = const$, $\psi = const$). The transformation of variables in accordance with (5.16) results in the mapping of parabolic surface $\tau = const$ (for arbitrary τ) onto the plane (ρ, φ_c). The line of this surface intersection with any paraboloid $\sigma = const$ is transformed into a circumference.

The case of two charged converging shells of angular width 2α located along the surface of a paraboloid is of specific interest since it demonstrates the placement of metal and slot antennas of finite width on surfaces of revolution. The two-dimensional problem in the form of a long line of two coaxial cylindrical shells corresponds to this case. In the parabolic line as in the conic line magnitudes C and W are constant. Hence the line is uniform and expressions (5.11) and (5.12) are valid for it. If the widths of the slot and the metal shell are same, the expression (5.14) is valid also. If the axis of the paraboloid is placed horizontally, then the metal radiator, as a similar radiator on a circular cone, is located at an acute angle to the ground, and that allows to increase the length of its arm and the antenna directivity at the same height of the supports.

The reciprocal comparison of the arms of different self-complementary radiators of the same height H with each other shows that the arm length of a flat radiator is equal to $S_f = H$. The arm length of a conic radiator depends on an angle θ_0 (see Fig. 5.2) and is equal to $S_c = H/\sin\theta_0$. In particular, if $\theta_0 = 30°$, $S_c = 2H$. The arm length of a parabolic radiator is calculated by a known formula

$$S_P = \sqrt{z_1(z_1 + p/2)} + p/2sh^{-1}\sqrt{2z_1/p}. \qquad (5.18)$$

Here (see Fig. 5.6) $z_1 = z + |z_0|$ is a length of a paraboloid from the vertex to an aperture, $|z_0| = -H^2/2(L + l)$ is the distance from a vertex to coordinates' origin (to a parabola focus), $p = (L + l)H^2/[2L(L + l) + H^2]$ is a parabolic parameter, $l = \sqrt{L^2 + H^2}$. In particular, if $\theta_0 = 30°$, $S_p = 2.204H$. This value is more by 10.2 per cent than S_c, including 7.7 per cent at the expense of increase in the structure length by $|z_0|$. If $\theta_0 = 15°$, $S_f = H$, as before. Magnitudes S_c and S_p are, respectively, 3.86H and 4.04H. Thus, a paraboloid has the maximum arm length.

For comparison, Fig. 5.7 shows the active R_A and reactive X_A components of the input impedance for a flat, conical and parabolic antenna, depending on the ratio λ/H of a wavelength to the height of the antenna. The angle at a cone vertex and the angle between straight lines connecting the focus of a paraboloid with aperture boundaries are equal to 30°. In Fig. 5.8 and 5.9 reflectivity ρ and standing wave ratio (SWR) of the same antennas are given depending on the ratio λ/H in a cable with a wave impedance 60 π Ohm (a) and 100 Ohm (b).

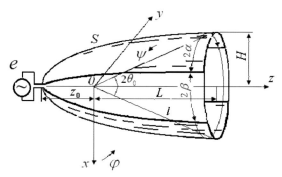

Fig. 5.6: Paraboloid of radiation.

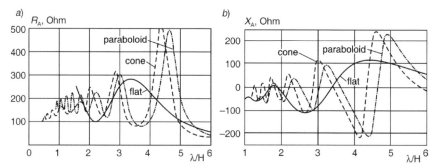

Fig. 5.7: Input impedance of a flat, conical and parabolic antenna, depending on the ratio λ/H of a wavelength to a height of an antenna.

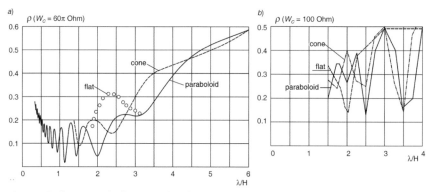

Fig. 5.8: Reflectivity of a flat, conical and parabolic antenna in a cable with a wave impedance of $60\,\pi$ Ohm (a) and 100 Ohm (b).

Calculations and experiments show that increase in the electrical length of an antenna generatrix improves characteristics of the radiator: matching with the cable, directivity and proximity of a main lobe of a directional pattern to a ground surface (see Table 5.1).

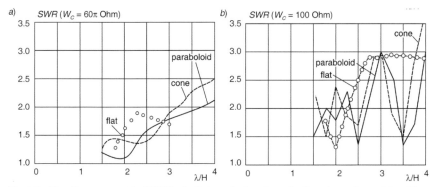

Fig. 5.9: Standing wave ratio (SWR) of a flat, conical and parabolic antenna in a cable with a wave impedance of 60 π Ohm (a) and 100 Ohm (b).

Table 5.1: Gain and efficiency of different radiators.

λ/H	Gain at θ_0					Efficiency at θ_0				
	flat	*cone*		*paraboloid*		*flat*	*cone*		*paraboloid*	
	90°	30°	15°	30°	15°	90°	30°	15°	30°	15°
1	2.8	16.1	17.8	11	12.4	0.97	0.96	0.98	0.98	0.98
2	1.9	4.7	4.9	4.2	6.8	0.91	0.96	0.98	0.96	0.97
2.5	2.3	3.4	3.4	3.5	5.3	0.92	0.98	0.95	0.95	0.97
3	1.9	2.5	2.5	2.8	4.4	0.93	0.90	0.91	0.95	0.95

5.4 Antenna on a pyramid edges

Analysis shows that the characteristics of antenna located along the edges of a horizontally lying pyramid are close to the characteristics of antenna located on the surfaces of a circular cone or paraboloid of radiation. Increase in the length of such antenna arms allows use of the antenna of the same height as the height of flat self-complementary radiator to improve the electrical characteristics or, at the same characteristics, to reduce its height. The simple shape of an antenna on pyramid sides in comparison with the forms of antennas located on the surfaces of rotation facilitates and simplifies the construction of large antennas for the medium-wave band (for example, for broadcasting). It is advisable to use them also in the short-wave range, especially on vehicles.

The previous sections consider flat and three-dimensional (volumetric) radiators that have a shape of self-complementary radiators but differ from them by finite dimensions. The inclined volumetric radiator of the same height can operate at lower frequencies than the vertical flat radiator. But in order to substantially reduce the operating frequency, it is necessary to increase the antenna dimensions and to simplify the design. The metal surfaces of the antennas are usually replaced by thin wire systems. First of all, it is important for three-dimensional antennas. Such structures for asymmetric versions of antennas with wires located on conical and parabolic surfaces are shown in Fig. 5.10.

Similar structures, whose wires are located along faces of regular (Fig. 5.11a) or irregular (Fig. 5.11b) pyramids with a horizontal axis make possible to simplify antenna designs. Taking into account a mirror image in the ground, one can consider that these radiators are symmetrical. An electric radiator (dipole) consists of two flat metal cones, located along the upper and lower pyramid edges. The magnetic radiator consists of two slots located along the side edges.

In order to simplify the construction, one can accomplish the metal radiator arm in the form of a flat triangle (Fig. 5.12). These antennas are similar to self-complementary antennas, but they do not concern them because their surfaces are not smooth surfaces of rotation. About such structures one can say that they do not have axial symmetry. Therefore, expressions (5.8) and (5.9) are incorrect here, since radiators are placed along the pyramid faces.

Their characteristics are close to the characteristics of volumetric self-complementary antennas. Since their electrical lengths are larger than those of flat antennas with the same height, they can be used for broadcasting at low and medium frequencies. For an approximate analysis of such antenna characteristics, primarily for calculating the wave and input impedances, one can use their proximity to the characteristics of an antenna located on a circular cone.

For calculating input impedances of these radiators, one can apply a previously used method based on calculating the wave impedance of a uniform line of infinite length. As in the case of wave impedance of the earlier considered conical line with a circular cross-section, the wave impedance of the pyramid is defined by angles at its vertex. This impedance depends on the ratio of magnitudes b and d, which is constant along the line having the shape of pyramid (b is the triangle width in the given cross-section and d is the distance between the plates). The wave impedance doesn't depend on absolute values of these magnitudes. This means that this long line is also uniform and its input impedance with increasing line length tends towards the wave impedance W.

Capacitance per unit length of this line in the first approximation is equal to the capacitance per unit length of a line from two metal flat cones with an angle 2γ at the vertex, which lie in the planes located at an angle 2δ to one another [33]. A flat cone differs from a triangular plate by that the base of an isosceles triangle is replaced by an arc of a circumference with point O as center (the edges of the flat cones are shown in Fig. 5.12 by a dotted line).

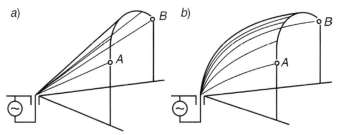

Fig. 5.10: Antennas with wires located on conical (a) and parabolic (b) surfaces.

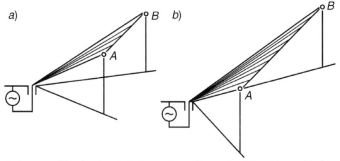

Fig. 5.11: Antennas with wires located along edges of a regular (a) and irregular (b) pyramids.

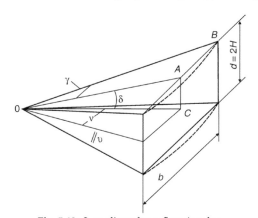

Fig. 5.12: Long line of two flat triangles.

In this case the capacitance C_l per unit length and the wave impedance W are calculated in accordance with 5.11, but k is equal to (see [33])

$$k = \left(\frac{1-\sin\gamma}{1+\sin\gamma}\right)^{\pi/2\delta}. \tag{5.19}$$

At that $\tan \gamma = AB/AO$, $\sin \delta = AC/AO$, $\tan \gamma / \sin \delta = AB/AC = b/d$. If the pyramid is regular, i.e., $b = d$, then $\tan \gamma / \sin \delta = 1$. The wave impedance W is a function of the ratio b/d and angle γ. If, for example, $b/d = 1$, i.e., dimensions of metal and slot radiators are equal, then an increase γ from 15° to 30° (and correspondingly an increase δ from 15.5° to 35.3°) leads to an increase of W from 42.3 to 45 Ohm. The input impedance of such radiators is smaller 60π, but it does not depend on the frequency, if the pyramid length tends towards infinity.

An electric radiator (dipole) consists of two flat metal cones, located along the upper and lower edges of the pyramid. Its input impedance Z_e is equal to impedance of the metal radiator, which is identical to the slot in shape and dimensions. If, for example, $b/d = 1$ and $\gamma = 15°$ (accordingly $\delta = 15.6°$), then $k = 0.0467$ and $Z_e = 42.2\,\pi$. The input impedance Z_s of the magnetic radiator is equal to input impedance of the metal radiator located next to the slot. In this case $\gamma = 15.6°$ and $\delta = 15°$, i.e., $k = 0.1923$ and $Z_s = 63\,\pi$. Accordingly, $Z_e Z_s = (51.6\,\pi)^2$. If $b/d = 2$, $\gamma = 30°$, then $\delta = 45°$, $k = 0.111$, $Z_e = 52.6\,\pi$. If $\delta = 30°$, then $\gamma = 45°$, $k = 0.00505$ and $Z_s = 28.2\,\pi$, i.e., $Z_e Z_s = (38.5\,\pi)^2$. As the calculations show, in the given case the product $Z_e Z_s$ depends on the ratio b/d, but it is always smaller than $(60\pi)^2$. The smaller angle γ, the closer wave impedance of the radiator located along the pyramid sides to the wave impedance of the conical radiator with the same angular width.

The electrical characteristics of symmetrical antennas located on a regular pyramid with $\gamma = \delta = 15°$ are shown by solid curves in Figs. 5.13 to 5.15, and also in Table 5.2, depending on the ratio λ/H, where λ is the wavelength, H is the antenna height. Each antenna arm is modeled as a thin triangular metal sheet with a height H and a thickness of $0.01H$. The gap between the arms at the pyramid vertex is $0.01H$. The antenna characteristics are calculated by means of the CST program. The active and reactive components of the input impedance for the antenna located along the edges of a regular pyramid with an angle 30° at the vertex are shown in Fig. 5.13. Reflectivity ρ and standing wave ratio (SWR) in cables with wave impedance $W_c = 60\pi$ (a) and 100 Ohm (b) are shown in Fig. 5.14 and 5.15 respectively. These figures show also the results of a conical antenna with the same angle at the vertex. One can see that the characteristics of these antennas are similar, but not identical. Here results of measuring an antenna mock-up located on the pyramid are given also. They confirm the calculations.

The electrical characteristics of an antenna located along the edges of an irregular pyramid are shown in Figs. 5.16 to 5.18 and in Table 5.2. The pyramid has a rectangular cross-section. The angular widths of a metal and a slot radiator are 60° and 30°, respectively. Here, for comparison, the characteristics of the antenna located on a pyramid with a square cross section are given. It can be seen from Fig. 5.18 that SWR of an antenna located

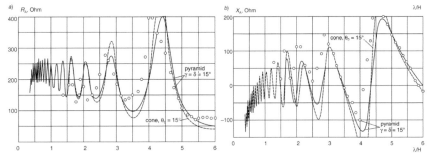

Fig. 5.13: Input impedance of an antenna on a regular pyramid and a conical antenna depending on the ratio λ/H.

Table 5.2: Gain and radiation efficiency of symmetrical antennas on pyramid's sides.

λ/H	Gain		Radiation efficiency	
	$\gamma = \delta = 15°$	$\gamma = 30°, \delta = 15°$	$\gamma = \delta = 15°$	$\gamma = 30°, \delta = 15°$
1	23.3	17.6	0.97	0.98
2	6.56	10.2	0.98	0.95
2.5	5.21	6.87	0.98	0.95
3	4.08	5.5	0.93	0.94

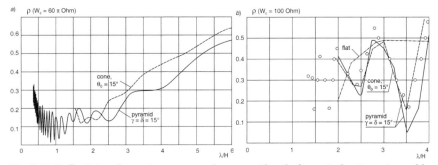

Fig. 5.14: Reflectivity of an antenna on regular pyramid and of a conical antenna in a cable with a wave impedance of 60 π Ohm (a) and 100 Ohm (b).

on an irregular pyramid is larger when using a coaxial cable with a wave impedance $W_c = 60\pi$ Ohm and is less when $W_c = 100$ Ohm. The results of these calculations show that properties of self-complementary radiators are not always preferable in comparison to properties of radiators with similar dimensions. As can be seen from Fig. 5.16, an increase of the angular width of the metal radiator increases the current along it and reduces active and reactive components of the input impedance, hence approximating its value to a wave impedance of a standard cable. If an asymmetric antenna uses

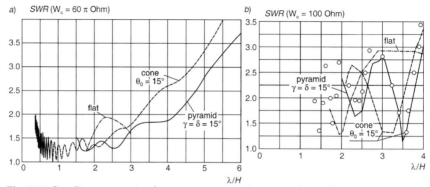

Fig. 5.15: Standing wave ratio of an antenna on regular pyramid and of a conical antenna in a cable with a wave impedance of $60\,\pi$ Ohm (a) and 100 Ohm (b).

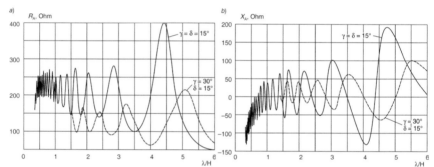

Fig. 5.16: Input impedance of antennas, located on a regular ($\gamma = \delta = 15°$) and irregular ($\gamma = 30°$, $\delta = 15°$) pyramid.

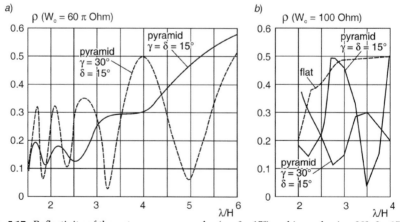

Fig. 5.17: Reflectivity of the antennas on a regular ($\gamma = \delta = 15°$) and irregular ($\gamma = 30°$, $\delta = 15°$) pyramid in cables with wave impedance $W_c = 60\pi$ (a) and $W_c = 100$ Ohm (b).

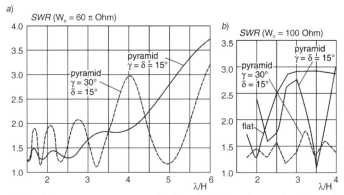

Fig. 5.18: SWR of the antennas on a regular ($\gamma = \delta = 15°$) and irregular ($\gamma = 30°$, $\delta = 15°$) pyramid in the cable with wave impedance $W_c = 60\pi$ (a) and $W_c = 100$ Ohm (b).

a special cable with a wave impedance $W_c = 30\pi = 94$ Ohm, the antenna located on the regular pyramid at the main frequency will be better matched to the cable. If a standard cable (with $W_c = 50$ Ohm) is used, an asymmetric antenna located along the edges of an irregular pyramid is better matched.

This result is easily explained. As is shown in Section 5.1, the wave impedance of the radiator located along the pyramid sides can be calculated in accordance with (5.11). On the other hand, this expression allows determination of the angular width of the antenna, which provides optimum matching of antenna with cable at a given wave impedance of this cable and a known vertical angle at the pyramid vertex. For example, if the cable wave impedance is 100 Ohm (in the case of symmetrical antenna version) and it is necessary to obtain the same magnitude of the antenna wave impedance, in accordance with the mentioned formula, the following relation must be fulfilled:

$$K\left(\sqrt{1-k^2}\right)/K(k) = 120\pi/100 = 3.770.$$

Function $K(k)$ for small values of an argument k is close to the value 1.57, i.e., $K\left(\sqrt{1-k^2}\right) = 5.919$. From the tables we find that $\sqrt{1-k^2} = 89{,}383°$, and $k = 0{,}01196$. If, for example, $\delta = 15°$, then from (5.19) $\gamma = 20.7°$. Similarly, in order to match a conic self-complementary antenna with a standard cable, whose wave impedance is smaller than the antenna wave impedance, one must increase the angular width of the metal radiator.

The typical characteristics of conic self-complementary antennas and of antennas located on the pyramid edges are presented in figures. In Fig. 5.19 the calculated curves and experimental magnitudes (circles) of R_A and X_A are given for conic antennas with different angles $2\theta_0$. In Fig. 5.20 *SWR* of the conic and flat antennas are compared with each other and demonstrated advantage of the conic antennas. In Fig. 5.21 pattern factors are presented for symmetrical antennas located along the edges of regular and irregular pyramid (in this case within the angular sector from 60° to 90°). The results of the calculation were checked on mock-ups placed in an anechoic chamber. A general view of the installation with an asymmetric antenna placed on the pyramid edges is shown in Fig. 5.22.

In conclusion of Chapter 5, we touch upon one more issue. In Chapters 2 to 4, the problems of extending the frequency range and increasing *PF* by including capacitive loads in linear metal radiators were considered. A similar method can be used in self-complementary antennas, including loads in the form of slots in metal radiators and in the form of metal strips of the same shape and dimensions in the slots, i.e., creating in-phase electric currents in metal radiators and in-phase magnetic currents in slots (see Fig. 5.23).

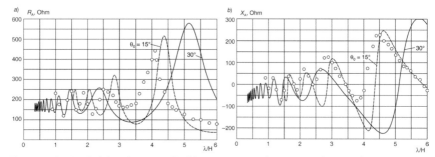

Fig. 5.19: Active R_A (a) and reactive X_A (b) components of input impedances of a symmetrical conic antennas with the angle 60° and 30° at vertex.

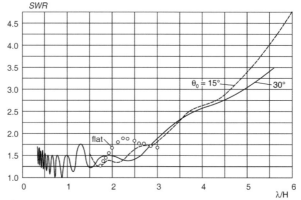

Fig. 5.20: Standing wave ratio of a symmetrical conic antennas with the angle 60° and 30° at vertex.

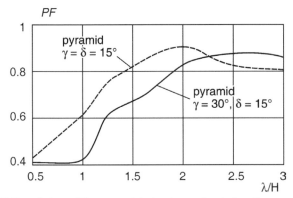

Fig. 5.21: Pattern factors of the symmetrical antennas located on the pyramid edges.

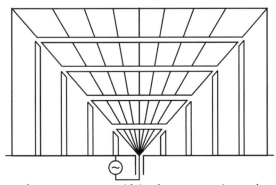

Fig. 5.22: General view of a measuring installation in anechoic chamber.

Fig. 5.23: Self-complementary antenna with in-phase currents in metal and slot radiators.

Self-complementary Antennas with Rotation Symmetry

6.1 Flat antennas

Antennas with rotational symmetry are a most complex variant of self-complementary antennas. They consist of several metal radiators and the same number of slots. Two antennas with rotational symmetry are shown in Fig. 6.1. Both have finite dimensions. Each antenna consists of four symmetrical radiators of the same size and shape: two metallic and two slots. In the first antenna (see Fig. 6.1a) the arm of each radiator has the form of a flat cone. This is the circular sector, the left and right sides of which coincide with the radii and the base with the circumference of the structure. In the second antenna (see Fig. 6.1b), the boundaries of the radiators do not coincide with radii. They have a log-periodic form. Antennas with radiators in the form of flat cones are simpler and their properties will be considered depending on the number of cones and variants of connecting the antenna with a generator or a cable. The properties of the antennas shown in Fig. 6.1b are similar, although there are additional details.

Let's consider variants of antennas with different numbers of symmetrical metal radiators (dipoles). Antennas with one metal dipole are presented in Fig. 6.2. They have different widths depending on the angle at the flat cone vertex: 135° (a), 90° (b), 45° (c), 22.5° (d). Antennas with two metal dipoles are given in Fig. 6.3. They differ from each other by a circuit of connection with the generator poles. The cones connected with each other by a thin line are connected to the same pole: adjacent cones in variant a are connected together; in variant b one cone is connected with the other cone through a cone, in variant c—three cones are connected with one

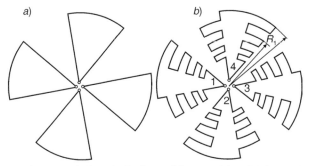

Fig. 6.1: Antenna from metal plates in the form of flat cone (a) and in log-periodic form (b).

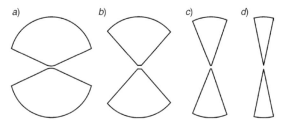

Fig. 6.2: Flat antennas with one metal radiator.

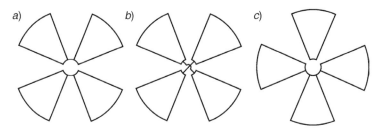

Fig. 6.3: Flat antennas with two metal radiators.

pole and the fourth cone is connected with another pole. The properties of diverse variants are different. In Fig. 6.4 different circuits of connection to the generator are demonstrated for an antenna with four metal dipoles. The main parameters of the considered antenna variants are given in Table 6.1. These are angle θ_0 between the axes of metal dipoles, the angular half-width α of a metal dipole and the angular half-width $\beta = \theta_0/2 - \alpha$ of a slot. The table gives a figure number for different antennas, the ratio C_l of the capacitance per unit length of each antenna to the absolute permittivity ε_0 of the air and also the antenna wave impedance W_1. The magnitude W_2 is the wave impedance of a three-dimensional antenna. About this, see further.

From this table it follows that the impedance of each antenna variant is constant in the first approximation, i.e., the current distribution along it is close to the current distribution along an equivalent long line. As already

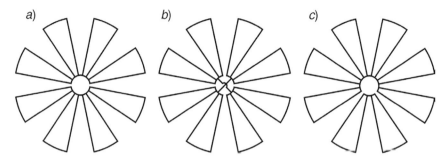

Fig. 6.4: Flat antennas with four metal radiators.

Table 6.1: Parameters and wave impedances of antennas with rotation symmetry.

Dipoles number	Fig.	θ_0	α	β	C_l/ε_0	W_1	W_2
1	6.2a	180°	67.5°	22.5°	2.94	40.8 π	40.8 π
	6.2b	180°	45°	45°	2	60 π	60 π
	6.2c	180°	22.5°	67.5°	1.35	88.7 π	88.7 π
	6.2d	90°	11.25°	78.75°	1.04	115.4 π	
2	6.3a	90°	22.5°	22.5°	4	17.8 π	20.4 π
	6.3b	90°	22.5°	22.5°	8	15 π	19 π
	6.3c	90°	22.5°	22.5°	5.35	22.4 π	26.5 π
4	6.4a	45°	11.25°	11.25°	12	9.9 π	8.4 π
	6.4b	45°	11.25°	11.25°	24	7.5 π	9.1 π
	6.4c	45°	11.25°	11.25°	13.04	10.6 π	10.6 π

said, the input impedance of a uniform line without losses with increasing its length tends towards its wave impedance. Similarly, the input impedance of the studied flat antenna changes: as the length of each antenna arm increases, its input impedance tends towards its wave impedance. If the antenna has a self-complementary structure, then its wave impedance is equal to 60π. As its length increases, the input impedance of the antenna tends towards 60π in accordance with expression (5.2).

In Fig. 6.5 curves show active and reactive components of input impedances for three antennas. They demonstrate how, with decreasing λ/L (λ is wavelength, L is arm length of the antenna), the value X_A approaches to zero and R_A to the wave impedance. Option 1 in Fig. 6.5 corresponds to an antenna with one metal radiator, option 2 corresponds to an antenna with two such radiators, and each pole of the generator is connected to two adjacent plates. Option 3 corresponds to the same antenna when the plates are connected to the generator pole through one.

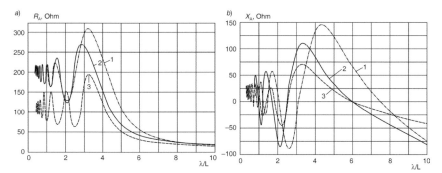

Fig. 6.5: Active (a) and reactive (b) components of input impedances of flat antennas.

As can be seen from Fig. 6.5, the input characteristics of antennas consisting of metal and slot radiators of the same shape and dimensions, if these antennas are located on a small metal sheet, are far from the characteristics of self-complementary antennas. The input impedance of the antenna shown in Fig. 5.1 slowly converges to the results, promised by expression (5.2). Similar results are obtained, if to compare input impedances of small-sized antennas with the values given in Table 6.1. The obvious closeness of the indicated magnitudes occurs when the structure dimensions are comparable with the wavelength. As an example, one can present results of using antennas with a single metal radiator in the high-frequency antenna array [37]. In these antennas, triangular metal surfaces are replaced by wires that diverge from the feed point—the lower vertex of the triangle (Fig. 5.1b). Each plate is replaced by five wires. The antenna size depends on the width of the frequency range: the wider the range, the longer the length H of the antenna. Let the required level of traveling-wave ratio (TWR) in the cable be greater than 0.6, and the frequency ratio equal to $k_f = 2$. Then the arm length H at the upper f_1, lower f_2 and central $f_3 = \sqrt{f_1 f_2}$ frequencies of the range is 0.3λ, 0.6λ and 0.42λ. If $k_f = 10$, the arm length at similar frequencies is 0.5λ, 4.5λ and 1.5λ. At the same time, multi-stepped transformers were used in the feed lines of antenna array, which ensure the required level of matching. This significantly alleviates the problem of selecting the wave impedances of feed lines.

As can be seen from the table, the described antennas have very different wave impedances, including small ones. Proximity of wave impedances of antennas and cables is a necessary condition for the antennas' efficiency. Known antenna variants do not satisfy this condition, since their wave impedances are substantially greater than the wave impedances of standard cables. Small wave impedances of new radiators should facilitate the task of matching and expanding the use of self-complementary antennas.

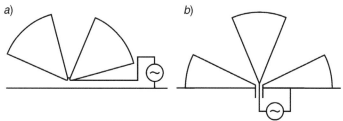

Fig. 6.6: Connecting asymmetric antennas to a generator.

In asymmetric structures, where the metal dipoles forming an antenna are replaced by monopoles located above the conducting surface (ground), the wave impedance is half as many. In Fig. 6.6 the circuits of connecting monopoles with generators are given. They are similar to the circuits shown in Fig. 6.3a and 6.3c for connecting dipoles.

6.2 Procedure of calculating flat self-complementary antennas

The method of calculating a simple self-complementary antenna with a single metal radiator was proposed by Carrel [33]. It is based on the conformal transformation of geometric figures. Transformation on a plane is called by conformal, if as a result the angles between any two intersecting lines remain unchanged and the lengths of all infinitesimal sections passing through a given point of the plane change in the same number of times. Under such a transformation, the capacitance preserves its magnitude. The method of conformal transformations in particular is used for calculating the capacitance of a system of wires by means of replacing the original structure with a structure whose capacitance is known.

Before applying this method, it is necessary to reduce the problem to a flat. Further, the problem is simplified if to move from a spherical coordinate system to a cylindrical one. When the coordinate system is changed in accordance with the uniqueness theorem, Laplace's equation must remain valid. As shown in [33], this condition is satisfied, if to go from a spherical coordinate system to a cylindrical, and the replaced variables are in accordance with equalities:

$$\rho = tg(\theta/2),\ \varphi_c = \varphi, \tag{6.1}$$

where ρ and φ_c are cylindrical coordinates, θ and φ are spherical.

The structure, considered in [33], in the shape of two arms, is shown in Fig. 6.7a. As a result of coordinate system's transformation in accordance with (6.1), each radial line becomes a point and the entire structure is transformed into two identical 'cuts' in the form of two sections of the horizontal line (see Fig. 6.7b). Further, the Schwarz-Christoffel's transform

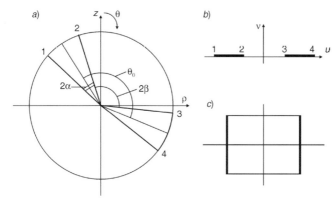

Fig. 6.7: Radiator of two flat metal cones (a), transition to two 'slits' (b), Schwarz-Christoffel's transformation (c).

maps this structure on to a rectangle, whose two opposite sides are plates of a flat capacitor having an infinite width. Capacitance of a capacitor per unit width is equal to

$$C_l = \varepsilon_0 K\left(\sqrt{1-k^2}\right)/K(k), \tag{6.2}$$

where $\varepsilon_0 = 10^{-9}/(36\,\pi)$ is the absolute permittivity of the surrounding medium, $K(k)$ and $K\left(\sqrt{1-k^2}\right)$ are complete elliptic integrals of the first kind from arguments k and $\sqrt{1-k^2}$, and k is equal to

$$k = \frac{\sin(\theta_0/2) - \sin\alpha}{\sin(\theta_0/2) + \sin\alpha}. \tag{6.3}$$

If the axial lines of arms (see Fig. 6.7a) coincide, then

$$k = \frac{1 - \sin\alpha}{1 + \sin\alpha}. \tag{6.4}$$

The basic variant considered in [33] is identical to the variant shown in Fig. 5.1b, where $\alpha = \beta = \pi/4$. In this case $k = 0.1716$, $K(k) = 0.5K\left(\sqrt{1-k^2}\right)$, i.e., according to (6.3), the capacitance of capacitor per unit width is equal to $C_l = 2\varepsilon_0$, and the antenna impedance is

$$W = 120\pi\,K(k)/K\left(\sqrt{1-k^2}\right) = 120\pi\varepsilon_0/C_l, \tag{6.5}$$

i.e., $W = 60\,\pi$.

In the case of two metal radiators (Fig. 6.8a) let's go in accordance with the conditions (6.1) from the spherical to the cylindrical coordinate system

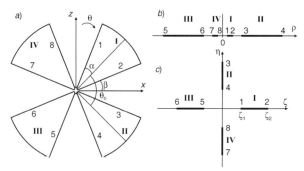

Fig. 6.8: Flat antenna of two metal radiators (a), transition to the cylindrical (b) and rectangular (c) coordinate system.

(Fig. 6.8b). As a transition results, the n-th radial line (the line numbers are indicated with Arabic numerals) goes to the point with the coordinate ρ_n. The point numbers coincide with the line numbers. The general expression for the magnitude ρ_n has the form

$$\rho_n = \tan\{[0{,}5[(2m-1)\,\theta_0/2 \pm \alpha]\}, \tag{6.6}$$

where $m = 1, 2, \ldots$ The sign '−' refers to the odd, and the sign '+' to the even values of n. Since $\alpha = \beta = \pi/8$, i.e., the slot and cone have the same width, then $\rho_n = \tan[(2n-1)\alpha/2]$. The general view of the transformed structure is given in Fig. 6.8b. This is a horizontal line with four slits ('cuts') connecting points with numbers $2n{-}1$ and $2n$.

As can be seen from the above, the described method allows to solve the problem for a metal radiator of arbitrary angular width with arbitrary angles between the axial lines of the arms. But in the case of two metal radiators, the described technique leads to four straight slits, in the case of N dipoles, to $2N$ slits. This makes possible to calculate several capacitances, but it does not allow bringing them together into one capacitance. Besides that, Fig. 6.8b does not take into account the relative position of elements. Therefore, in order to consider such a structure of several metal radiators, it is necessary to transit to rectangular coordinate system (Fig. 6.8c). For this aim, one can use the expression

$$\varsigma_n = \sqrt{\tan(2\tan^{-1}\rho_n)}. \tag{6.7}$$

The antenna consists of four identical flat metal cones evenly spaced along the circumference, i.e., separated by the same slots. The 'slit' numbers are shown in the figure in Roman numerals. The total angular width of the cone and the adjacent slot is equal to $\theta_0 = 2(\alpha + \beta) = \pi/2$. In accordance with (6.7) $\varsigma_1 = 0.6436$, $\varsigma_2 = 1.5538$, other coordinates are similar.

As already mentioned, the generator poles can be connected to different elements. In Fig. 6.9 there are three connection options. The cones connected

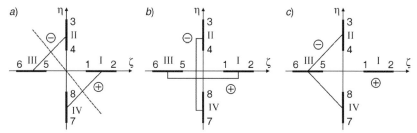

Fig. 6.9: Three options of antenna connection to generator poles: two neighboring cones (a) and two cones through one (b) are connected to each pole, three cones are connected to first pole and one to second (c).

with each other by a thin line are connected to the same generator pole. One pole is marked with a sign '+', the other with a sign '−'. In the first option (Fig. 6.9a) two neighboring metal cones are connected to each pole, in the second option (Fig. 6.9b) the cones are connected to each pole through a cone. In the third option (Fig. 6.9c), three cones are connected to one pole and one cone to another pole.

The dashed line in Fig. 6.9a is the symmetry axis of the structure. This is zero potential line. The capacitance between the poles on both sides of the line consists of two pairs of different capacitances. The capacitances of the first pair are formed by flat cones with coincident axes. They are calculated in accordance with expressions (6.4) and (6.2), and each one is equal to $C_{11} = 1.36\varepsilon_0$. Capacitances of the second pair are capacitances between the cones with mutually perpendicular axes. As shown in [24], the value of such capacitance can be obtained by means of a quadrant shown in Fig. 6.10a. This value is twice as large as calculated by formula (6.2), if k is equal to

$$k = \sqrt{\frac{(a^2 + c^2)(b^2 + d^2)}{(a^2 + d^2)(b^2 + c^2)}}.$$

Here a, b, c, d are the coordinates of the 'slits' ends in Fig. 6.10a. In our case, $a = c, b = d$, i.e.,

$$k = 2ab/(a^2 + b^2). \tag{6.8}$$

In accordance with (6.7) and (6.8) $a = \varsigma_1$, $b = \varsigma_2$, $k = 0.7071$, i.e., $K(k) = K\left(\sqrt{1-k^2}\right) = 1.8541$, and the capacitance between the cones with mutually perpendicular axes in Fig. 6.10a is equal to $C_{12} = 2\varepsilon_0$. The total input capacitance of the first antenna variant is $C_1 = 2(C_{11} + C_{12})\varepsilon_0 = 6.73\varepsilon_0$.

It is not difficult to be convinced that in the second variant of connecting metal cones with the generator poles, the capacitance consists of four elements C_{12}, i.e., $C_2 = 8\varepsilon_0$. Thus, the wave impedances of the first and second variants are 17.8 π and 15 π, respectively. In the third variant, the capacitance between one metal cone and three others consists of two parts. The first of these is the capacitance between a single cone and two side ones of the triple. It is equal, as is not hard to see, $C_{31} = 4\varepsilon_0$. The capacitance between two identical radiators is equal to $C_{11} = 1.36\varepsilon_0$, i.e., the total capacitance is $C_3 = 5.36\varepsilon_0$ and the wave impedance is $W = 22.4\pi$.

As already said, proximity of wave impedances of antennas and cables is a necessary condition for the antennas' efficiency. To accomplish this condition, one must change the width of metal radiators. If to make in the case of the first option connecting four metal cones with generator the angle $\alpha = 14°$, and $\beta = 31°$, then, as calculations show, $W \approx 32\pi$, and the wave impedances of a similar asymmetric antenna, shown in Fig. 6.6a, is equal to $W \approx 16\,\pi \approx 50$ Ohm (see Fig. 12.5b).

Calculating wave impedances of antennas from four metal radiators can be carried out in a similar manner. As shown in Fig. 6.4, there are three possible options for connecting flat cones to the poles of the generator. In the first variant of connection, the total capacitance consists of four capacitances $C_{11} = 1.04\varepsilon_0$ formed by flat cones with coincident axes (cone width is 22.5°) and four capacitances $C_{12} = 2\varepsilon_0$ between flat cones with mutually perpendicular axes, located in two quadrants (two capacitances in either quadrant, see Fig. 6.10b). The total capacitance is $C_1 = 12.2\varepsilon_0$, the wave impedance is $W = 9.9\,\pi$. In the second variant, the capacitance consists of eight elements $C_{12} = 2\varepsilon_0$ between flat cones with mutually perpendicular axes, located in four mentioned quadrants. The total capacitance is $C_2 = 16\varepsilon_0$, the wave impedance is $W = 7.5\,\pi$. In the third variant, the capacitance C_3 consists of two capacitances C_{11}, two capacitances C_{12} and four capacitances between flat cones, located at different angles to each other; total $C_3 = 11\varepsilon_0$. The wave impedance is $W = 10.6\,\pi$.

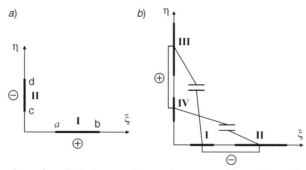

Fig. 6.10: Quadrants for calculating capacitance of antennas with two (a) and four (b) dipoles.

6.3 Three-dimensional antennas with rotation symmetry

A step forward with respect to simple vertical antennas in the form of a flat metal dipole with a plate of angular width equal to a slot angular width was made as a result of creating the flat self-complementary antennas with rotation symmetry. They are located in a vertical plane and an axis of their circular symmetry coincides with a perpendicular to this plane, i.e., it is horizontal. Another step forward is the development of three-dimensional (volumetric) antennas. Their appearance became possible when it was proved that the antenna based on the complementarity principle can be placed not only on the plane, but also on the surface of rotation; for example, on the surface of a circular cone or paraboloid located around the horizontal axis.

Comparison of conic and cylindrical problems in [33] showed that one problem reduces to another if replaced variables satisfy conditions (6.1) that were formulated for conic and cylindrical coordinate systems with the coincident vertical axes. Comparison of these parabolic and cylindrical problems shows [35] that their equivalence takes place, if the variables satisfy the conditions

$$\rho = \sigma, \varphi_c = \varphi, \tag{6.9}$$

where ρ and φ_c are cylindrical coordinates, σ and φ are parabolic.

In the conic problem (Fig. 6.11a) an exciting emf is located at a cone vertex and radiator arms are located along its surface. Boundaries of each arm coincide with cone generatrices. Angular lengths of arcs along circular conical aperture, corresponding to the metal radiator and the slot, are also designated as 2α and 2β. A paraboloid (Fig. 6.11b) has a similar aperture shape.

A rigorous analysis (see [32]) shows that the input impedances of the metal and slot radiators located on an infinite circular cone or paraboloid are

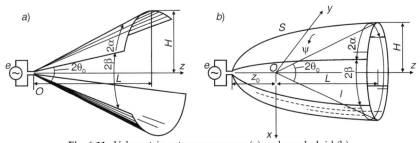

Fig. 6.11: Volumetric antennas on cone (a) and paraboloid (b).

connected with each other by the expression (5.1). For $\alpha = \beta$ the expression (5.2) is valid. This is natural, since their structures can be considered as two-wire long lines and these lines are uniform, i.e., in this respect, they do not differ from the corresponding long line located on the surface of a circular cylinder with a constant radius of cross section. Therefore, if $\alpha = \beta$, then the expressions (6.2), (6.4) and (6.5) are valid for conical and similar parabolic antennas. If $\alpha \neq \beta$, then k is calculated in accordance with (6.3).

As shown in Section 5.1, the conic structure with one symmetrical slot has properties similar to a flat structure with a similar slot. Really, in general case the relation analogous to expression (5.2) is true for a volumetric structure with arbitrary number of symmetrical slots, located at a circular metal cone or paraboloid of infinite length and excited on its vertex.

In conical problem not only axes of different cones coincide with a common symmetry axis, but the vertexes of these cones also coincide with each other. Unlike this, in the parabolic problem, the vertexes of different paraboloids are shifted along the common axis, so that each paraboloid is located separately. Capacitance per unit length measured along the axis of the cone and paraboloid, and also wave impedance of each antenna, are determined by the same formulas. But, if the apertures of the cone and the paraboloid are identical, i.e., their heights are equal to each other, the length of a parabolic generatrix is more than the length of a conic generatrix and accordingly, an effective height of the parabolic antenna is more. And, of course, effective heights and side lengths of both volumetric antennas are much more than the effective height and length of a side boundary of the flat antenna with the same geometric height. Therefore, they have for a given frequency and height the higher effectiveness in comparison with a flat antenna.

In Fig. 5.8 and 5.9 reflectivity ρ and *SWR* of symmetrical antennas located along surfaces of the circular paraboloid, the circular cone and the vertical flat antenna are presented. The angle $2\theta_0$ at cone vertex and the angle $2\theta_0$ between opposite focal radius-vectors of the paraboloid are equal to 30°. The wave impedances of cables are equal to $60\,\pi$ Ohm. These graphs allow comparison of electrical characteristics of antennas with each other. They confirm the advantage of parabolic and conic antennas over flat antenna in a predetermined frequency range.

Three-dimensional antennas, as well as flat antennas, can consist of several metal and slot radiators connected in parallel to each other. They can also be named antennas with rotational symmetry. An example of such conical antenna is given in Fig. 6.12a. This antenna, as well as the flat antenna, shown in Fig. 6.1a, consists of two metal dipoles. When calculating its characteristics, first, using conditions (5.3), we go to the plane problem. Since all points of each its generatrix are located at the same angle θ to the axis z, in consequence of the transformation they will fall into a common point and the metal plate will become a section of the circumference (Fig. 6.13b). If the boards have the same width and are separated by slots of the same width, then the coordinates of the initial (x_{m1}, y_{m1}) and final (x_{m2}, y_{m2}) points of these sections are determined by the expressions

$$x_{m1(2)} = \rho\cos\varphi_{m1(2)}, \ y_{m1(2)} = \rho\sin\varphi_{m1(2)}, \tag{6.10}$$

where, in accordance with (6.1), the quantity ρ is $\rho = tg(\theta/2)$ ($\theta/2$ is half angle at the cone vertex) and φ_m is the angle φ measured along the circumference from the axis x to the point m. It is equal to

$$\varphi_m = (2m-1)\varphi_0/2 \pm \alpha,$$

where magnitudes of angles φ_0 and α are clear from Fig. 6.12, signs '–' and '+' correspond to the beginning and end of the section.

As in the case of similar flat antennas, three options are possible for connecting different antenna elements to the generator poles (Fig. 6.13). The boards connected to each other by a thin line are connected to the same generator pole. One pole is marked with a sign '+', the other with a sign '–'. In the first (Fig. 6.13a) and the second (Fig. 6.13b) option, two metal boards are connected to each pole: in the first case they are neighboring boards,

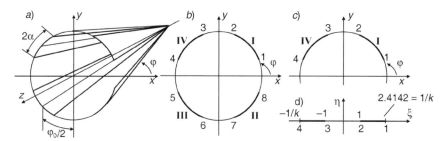

Fig. 6.12: Conical antenna with two metal radiators: a—general form, b—transition to a plane problem, c—symmetry division, d—conformal mapping.

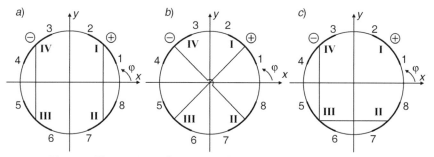

Fig. 6.13: Three options of connecting the antenna to the generator poles.

in the second, the boards are connected via a board. In the third option (Fig. 6.13c) one board is connected to one pole, and three to the other.

The axis x in Fig. 6.13a and 6.13b is the axis of system symmetry. This is zero potential line. The antenna capacitance between the poles with both sides of the line consist of two pairs of different capacitances. The capacitances of first pair are formed by sections of a circumference (arcs) with coinciding symmetry axes (see Fig. 6.13). They are calculated in accordance with (6.4) and (6.2), and capacitance is equal to $C_{11} = 1.364\varepsilon_0$.

The capacitances of second pair are capacitances between arcs located in neighboring quadrants (see Fig. 6.12c). Each arc is the circumference section of a radius ρ. If to introduce the function $z = x + jy$, where $x = \rho\cos\varphi, y = j\rho\sin\varphi$, then the value z at each point of the arc is equal to $z = \rho(\cos\varphi + j\sin\varphi)$. Let angular widths of the metal board and the slot be same. Then arc I is located between points with coordinates $x_1 = \rho\cos\varphi_1, y_1 = \rho\sin\varphi_1$, and $x_2 = \rho\cos\varphi_2, y_2 = \rho\sin\varphi_2$, where $\varphi_1 = 22.5°, \varphi_2 = 67.5°$. Using conformal mapping in the form

$$\varsigma = jz/\rho + \rho/(jz), \qquad (6.11)$$

where $\varsigma = \xi + j\eta, z = x + jy$, we obtain

$$\varsigma = j(\cos\varphi + j\sin\varphi) - j/(\cos\varphi + j\sin\varphi) = 2\sin\varphi, \qquad (6.12)$$

i.e., in the plane ς the arc section becomes the axis section between the points $\xi_1 = 2\sin\varphi_1 = 0.765$ and $\xi_2 = 2\sin\varphi_2 = 1.848$ (Fig. 6.12d). Similarly, the second arc section is transformed—between points 3 and 4. Dividing the points coordinates to ξ_2, we obtain the values given in Fig. 6.12d, from which it follows that in this case $k = 0{,}41$. Using the Schwarz-Christoffel's map, we find, in accordance with (6.2), that the capacitance of a structure shown in Fig. 6.12c is equal to $C_{12} = 1.583\varepsilon_0$.

Accordingly, the total antenna capacitance for the first variant of its connection to the generator poles is equal to $C_1 = 2C_{11} + 2C_{12} = 5.9\varepsilon_0$ and the wave impedance is $W = 20.4\,\pi$ Ohm.

The capacitance of the second antenna variant consists of four capacitances C_{12} between arcs located in neighboring quadrants (Fig. 5.16b) and is equal to $C_2 = 6.3\varepsilon_0$. This means that the wave impedance of the second variant is $W = 19\,\pi$ Ohm. In the third variant, the capacitance between one arc and three others consists of two parts. The first part is the capacitance between a single arc and two side arcs of the triple. It is equal, as is not hard to be convinced, to $2C_{12} = 3.17\varepsilon_0$. The capacitance between two arcs with coinciding axes is equal to $C_{11} = 1.364\varepsilon_0$, i.e., the total power is equal to $C_3 = 4.53\varepsilon_0$ and the wave impedance is $W = 26.5\,\pi$ Ohm.

Calculation of wave impedances of antennas from four metal radiators can be performed in a similar way. As shown in Fig. 6.4, here one can also separate three main options of connecting boards to the generator poles.

In the first and second connection option, the capacitances of four pairs of arcs with coinciding symmetry axes are included in full capacitance (each capacitance is equal to $C_{11} = 1.04\varepsilon_0$), in the third variant it is necessary to take into account the capacitances of two such pairs. The calculation is complicated by the need to take into account mutual capacitances not only between neighboring, but also remote arcs. This circumstance can be taken into account by increasing the angles φ_1 and φ_2 respectively, but this solution has an approximate character.

The calculation gives the following results. The total capacitance of the first connection variant is equal to $C_1 = 14.3\varepsilon_0$, the second—$C_2 = 11.2\varepsilon_0$, the third—$C_3 = 11.3\varepsilon_0$. The wave impedance of the first variant is $W = 8.4\,\pi$ Ohm, of the second—$W = 9.1\pi$ Ohm and of the third—$W = 10.6\pi$ Ohm.

The wave impedances W_2 of three-dimensional antennas located on conical and parabolic surfaces of rotation are given in Table 6.1 and allow comparing them with the wave impedances W_1 of flat antennas. They are the same in antennas with one metal radiator and in antennas with two and four, such radiators are a little more.

The obtained results show that the class of self-complementary antennas is much wider than what was previously thought. This class should be supplemented, firstly, by structures consisting of several metal and slot radiators, and, secondly, by three-dimensional structures located on the surfaces of rotation, in particular on the surfaces of a circular cone or paraboloid. Comparison of cylindrical, conical and parabolic problems allows to determine the relationship between variables that ensure their mathematical equivalence.

The proximity of the values of wave impedances of cables and antennas is a necessary condition for the efficiency of antennas. The known variants of self-complementary antennas do not satisfy this condition, since their wave impedances are greater than the wave impedances of standard cables. The antennas considered here have different wave impedances, including a significantly smaller one. This makes it much easier to reconcile in a wide frequency range and expand the scope of self-complementary antennas. Three-dimensional radiators increase substantially the efficiency of the antennas of the same height.

PART 2
MULTI-FREQUENCY ANTENNAS

7

Multi-wire Structures Parallel to Metal Surface

7.1 Related long lines parallel to the ground

Question of utility of multifrequency antennas and fundamental similarity of their employment with employment of wide-band antennas has been considered in the Introduction. Of course, these arguments are valid if the multifrequency properties actually take place. But the necessity to create a calculation technique used subsequently for the analysis of multi-frequency antennas was caused by other reasons. An application of long and medium waves in the first stage of an existence of radio communication and broadcasting led to the widespread use of *L*-shaped (inverted *L*) antennas with a developed horizontal load and to the need to determine the magnitude of the load, depending on its circuit and suspension height. An example of such a load consisting of two parallel wires located above an infinite metal surface (above the ground) is shown in Fig. 7.1. In general case, the number of wires can be substantially greater than two. They form a structure of electrically connected lines located above the ground. The theory of coupled lines [38], developed by Pistolcors, became the basis for calculating structures from parallel wires. This theory allowed to determine the characteristics of different antennas and to analyze the processes in multi-wire cables.

Naturally, the basis for the proposed method of calculation was the method of calculating a two-wire long line, the so-called telegraph equations. These equations for the current and potential of each wire *i* of the line have the form

$$-\frac{\partial u_1}{\partial z} = jX_{11}i_1 + jX_{12}i_2, \quad u_1 = j\frac{1}{k^2}\left(X_{11}\frac{\partial i_1}{\partial z} + X_{12}\frac{\partial i_2}{\partial z}\right),$$

$$-\frac{\partial u_2}{\partial z} = jX_{22}i_2 + jX_{12}i_1, \quad u_2 = j\frac{1}{k^2}\left(X_{22}\frac{\partial i_2}{\partial z} + X_{12}\frac{\partial i_1}{\partial z}\right). \quad (7.1)$$

Here u_i is the potential of the wire i with respect to ground, i—the current along the wire i, $X_{ik} = \omega\Lambda_{ik}$—the self or mutual inductance per unit length of the wire.

Two left equations of the set (7.1) are based on the fact that a potential drop on an element dz of each wire is the result of emf influence. Emfs are excited by their own currents and currents of neighboring wires. Two other equations are written on the basis of electrostatic equations connecting charges and potentials in accordance with the continuity equation.

Dependence of a current on the z coordinate is taken in the form $\exp(\gamma z)$, where γ is the propagation constant. Differentiation of the right-hand equations and their substitution into the left-hand ones reduces them to a set of uniform equations, which show that the propagation constant in the system of two metallic wires is k. We seek a solution of equations in the form

$$u_1 = A \cos kz + jB \sin kz.$$

Assuming $z = 0$ to determine the constant values A and B, we obtain

$$i_{1(2)} = I_{1(2)} \cos kz + j\left[\frac{U_{1(2)}}{W_{11(22)}} - \frac{U_{2(1)}}{W_{12(21)}}\right]\sin kz, \quad u_{1(2)} = U_{1(2)} \cos kz + j\sum_{s=1}^{2}\rho_{1(2)s}I_s \sin kz,$$

$$(7.2)$$

where $I_{1(2)}$ and $U_{1(2)}$ are the current and the potential at the beginning of the wire 1 or 2 (at the point $z = 0$), $W_{1(2)s}$ and $\rho_{1(2)s}$ are the electrostatic and electrodynamic wave impedance between wires 1 or 2 and wire s.

In the general case, when the system consists of N parallel metal wires located above the ground, the expressions for the current and potential of the wire n have the form

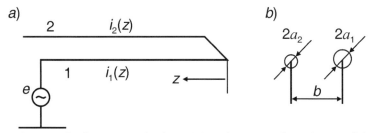

Fig. 7.1: An example of an antenna load consisting of two wires located in parallel to the infinite metal surface: a—a circuit, b—a cross section.

$$i_n = I_n \cos kz + j\left[\frac{2U_n}{W_{nn}} - \sum_{s=1}^{N}\frac{U_s}{W_{ns}}\right]\sin kz, u_n = U_n \cos kz + j\sum_{s=1}^{N}\rho_{ns}I_s \sin kz, \text{(7.3)}$$

where both I_n and U_n are the current and the potential at the beginning of a wire n (at the point $z = 0$), W_{ns} and ρ_{ns} are the electrostatic and electrodynamic wave impedance between the wires n and s, respectively:

$$\rho_{ns} = \frac{\alpha_{ns}}{c}, \quad W_{ns} = \begin{cases} 1/(c\beta_{ns}), n = s, \\ -1/(c\beta_{ns}), n \neq s. \end{cases} \quad \text{(7.4)}$$

Here α_{ns} is a potential coefficient (taking into account the mirror image in the perfectly conductive surface of the ground), β_{ns} is a coefficient of electrostatic induction, c is the speed of light. The coefficients α_{ns} and β_{ns} are related by expression

$$\beta_{ns} = \Delta_{ns}/\Delta_{N}, \quad \text{(7.5)}$$

where $\Delta_N = |\alpha_{ns}|$ is a determinant $N \times N$, and Δ_{ns} is an algebraic complement of the determinant Δ_N.

For an asymmetric line of two wires, one can write

$$\frac{1}{W_{11}} = \frac{\rho_{22}}{\rho_{11}\rho_{22} - \rho_{12}^2}, \quad \frac{1}{W_{22}} = \frac{\rho_{11}}{\rho_{11}\rho_{22} - \rho_{12}^2}, \quad \frac{1}{W_{12}} = \frac{\rho_{12}}{\rho_{11}\rho_{22} - \rho_{12}^2}, \quad \text{(7.6)}$$

i.e.,

$$\left(\frac{1}{W_{12}} - \frac{1}{W_{11}}\right):\left(\frac{1}{W_{12}} - \frac{1}{W_{22}}\right) = \frac{\rho_{22} - \rho_{12}}{\rho_{11} - \rho_{12}} = g. \quad \text{(7.7)}$$

Here g is a some constant.

Finally, if the wires of the asymmetric line have unequal lengths, or if concentrated loads are included in them, it is necessary to divide the line into sections. The expressions for the current and potential in the section m of the wire n take the form

$$i_n^{(m)} = I_n^{(m)} \cos kz_m + j\left(\frac{2U_n^{(m)}}{W_{nn}^{(m)}} - \sum_{s=1}^{N}\frac{U_s^{(m)}}{W_{ns}^{(m)}}\right)\sin kz_m,$$

$$u_n^{(m)} = U_n^{(m)} \cos kz_m + j\sum_{s=1}^{N}\rho_{ns}^{(m)} I_s^{(m)} \sin kz_m, \quad \text{(7.8)}$$

where $I_n^{(m)}$ and $U_n^{(m)}$ are a current and potential of a wire n at a beginning of a section m (at the point $z_m = 0$), respectively, N is a number of wires in the section m, $W_{ns}^{(m)}$ and $\rho_{ns}^{(m)}$ are electrostatic and electrodynamic impedances between the wires n and s in the section m. Equations (7.8) generalize equations (7.3).

In order to solve each system of equations, boundary conditions are used. They establish the absence of currents at the free ends of the wires, continuity of current and potential along each wire, sudden changes in potential at the points of inclusion of loads and generator e. Calculating the magnitude of the current $J(0)$ at the feed point, one can find the input impedance of an asymmetric line

$$Z_l = e/J(0). \tag{7.9}$$

It is approximately equal to the reactive input impedance of an antenna whose equivalent is the asymmetrical line. The input impedance of the antenna can be found more precisely if the antenna is treated as a linear radiator, the current along which is equal to the total current of the line.

When calculating the input impedance of an antenna, it is necessary as a rule to find a field E_ς on the antenna surface. And one must have in mind that while the function $J(\varsigma)$ of a current is continuous along the entire length of the antenna and is sinusoidal at each section, the function $dJ/d\varsigma$ may experience jumps on the sections' boundaries.

Equations (7.8) use the wave impedances $W_{ns}^{(m)}$ and $\rho_{ns}^{(m)}$; equations (7.3) use similar quantities. Magnitudes of the wave impedances, as seen from (7.4), are determined by the potential coefficients. These coefficients are calculated by a method of average potentials in accordance with an actual arrangement of antenna wires. The method of How is a simplest version of a such methods. It is easy to show that the mutual potential coefficient of two parallel wires with equal length, dimensions and location of which are given in Fig. 7.2, is equal to

$$\alpha_{ns} = \alpha(L,l,b)/(2\pi\varepsilon), \tag{7.10}$$

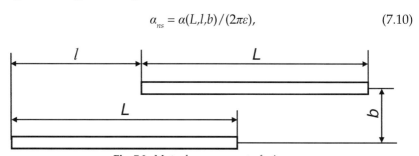

Fig. 7.2: Mutual arrangement of wires.

where

$$\alpha(L,l,b) = \frac{1}{2L}\left[(L+l)sh^{-1}\frac{L+l}{b} + (L-l)sh^{-1}\frac{L-l}{b} - 2Lsh^{-1}\frac{l}{b} - \sqrt{(L+l)^2 + b^2} - \sqrt{(L-l)^2 + b^2} + 2\sqrt{l^2 + b^2} \right],$$

i.e., $\rho_{ns} = \alpha_{ns}/c = \alpha(L,l,b)/(2\pi\varepsilon_0,\varepsilon_r c) = 60\alpha(L,l,b)\varepsilon_r$.

7.2 Meandering loads of wire antennas

An example of using a wire system located in parallel to an infinite metal surface can serve a meandering wire structure applicated as a horizontal load of an inverted L antenna. Analysis of such a load may be performed by means of the theory of coupled lines. As is known, an antenna load serves to increase an effective height by changing the current distribution along the radiating vertical antenna wire. The equivalent length of the horizontal section of the antenna is equal to

$$l_e = \frac{1}{k}\tan^{-1}\left(\frac{W_H}{W_V}\cot kL \right) \approx \frac{W_V l}{W_H}. \tag{7.11}$$

If dimensions of this section are small compared to a wavelength, then its equivalent length is inversely proportional to the wave impedance W_H of the horizontal section and is directly proportional to its length l, as well as the wave impedance W_V of the vertical section. Therefore, the horizontal load is usually made in the form of several parallel wires, whose ends are connected to each other (Fig. 7.3a).

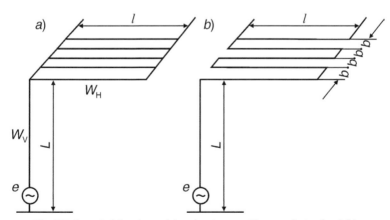

Fig. 7.3: Inverted L antenna (a) and antenna with meandering load (b).

In Fig. 7.3b the antenna circuit is given where wires of the horizontal load are connected to each other in series [39]. In this circuit, a path of the current along the load is lengthened, and its equivalent length is increased. An increase in the equivalent length of the load makes it possible to increase the effective height of the antenna. As can be seen from the above, in this case the structure of the wires is used not to create an antenna with several series resonances and, accordingly, a good matching level at several operating frequencies, but to obtain a qualitative communication at one frequency. It should remind that inverted L antennas are widely used on ships of the marine fleet, and there the problem of increasing an altitude and accordingly an effective height of the antenna is acute.

As already mentioned, in order to calculate the input impedance of the load, one can use the theory of coupled parallel long lines located above the infinite metal surface. The current and potential of each wire of an asymmetrical line consisting of N parallel wires are determined by expressions (7.3), and the wave impedances between the wires are determined by expressions (7.4). For example, determining the input impedance of three wires load, shown in Fig. 7.4, begins with writing the boundary conditions for the currents and potentials:

$$u_1(0) = u_2(0), \quad i_1(0) + i_2(0) = 0, \quad i_3(0) = 0,$$
$$u_2(l) = u_3(l). \quad i_2(l) + i_3(l) = 0, \quad u_1(l) = e. \tag{7.12}$$

It is supposed that the radii of the wires are small compared to the distances b between them. Since the height L of their suspension is commensurable with the length of the wire l, the method of average potentials is used to determine the potential coefficients. Taking into account the mirror image, we get:

$$\alpha_{11} = \alpha_{22} = \alpha_{33} = \alpha_1/(2\pi\varepsilon_0), \tag{7.13}$$

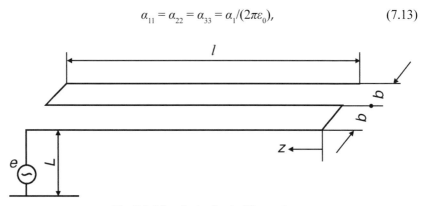

Fig. 7.4: Meandering load of three wires.

where

$$\alpha_1 = \ln(l/a) - 0.307 - sh^{-1}(l/2L) + \sqrt{1+(2L/l)^2} - 2L/l.$$

Similarly

$$\alpha_{12} = \alpha_{21} = \alpha_{23} = \alpha_{32} = \alpha_2/(2\pi\varepsilon_0), \ \alpha_{13} = \alpha_{31} = \alpha_3/(2\pi\varepsilon_0), \tag{7.14}$$

where

$$\alpha_2 = sh^{-1}(l/b) - \sqrt{1+(b/l)^2} + b/l - sh^{-1}\left(l/\sqrt{4L^2+b^2}\right) + \sqrt{1+\left(4L^2+b^2/l^2\right)} - \sqrt{4L^2+b^2/l^2},$$

and α_3 turns out from α_2 if we replace b by $2b$. Since the distance between neighboring wires of the load is small in comparison with L and l, we will put, for simplicity of calculations that

$$\alpha_2 = \alpha_3. \tag{7.15}$$

This assumption does not cause a significant error, because $sh^{-1}x = \ln\left(x+\sqrt{x^2+1}\right)$, i.e., the error is logarithmic. It is justified also because the length of the connecting bridges at the ends of the load can be taken into account in the used method only indirectly (by replacing l by $l+b$).

From (7.4) and (7.13)–(7.15) we obtain:

$$\rho_{ns} = \begin{cases} \rho_1, n=s, \\ \rho_2, n \neq s, \end{cases} \tag{7.16}$$

where $\rho_1 = 60\alpha_1$, $\rho_2 = 60\alpha_2$. The coefficients β_{ns} and α_{ns} are related by (7.5). From (7.4), (7.5) and (7.16) it is follows that

$$W_{ns} = \begin{cases} W_1 = 1/(c\beta_{11}), n=s, \\ W_2 = -1(c\beta_{12}), n \neq s, \end{cases} \tag{7.17}$$

and for three-wire load

$$\frac{1}{W_1} = \frac{\rho_1+\rho_2}{(\rho_1-\rho_2)(\rho_1+2\rho_2)}, \frac{1}{W_2} = \frac{\rho_2}{(\rho_1-\rho_2)(\rho_1+2\rho_2)}. \tag{7.18}$$

Substituting expressions (7.3) for the current and potential into the boundary conditions (7.12) and taking into account (7.16) and (7.18), we obtain:

$$U_2 = U_1, I_2 = -I_1, I_3 = 0, U_3 = U_1 + j(\rho_2-\rho_1)I_1 \tan kL,$$

$$I_1 = j\frac{2U_1 \tan kl}{\rho_1 + 2\rho_2 - \rho_1 \tan^2 kl}, U_1 = e\frac{\rho_1 + 2\rho_2 - \rho_1 \tan^2 kl}{\cos kl[\rho_1 + 2\rho_2 - (3\rho_1 - 2\rho_2)\tan^2 kl]}. \quad (7.19)$$

From here the input impedance of three-wire load

$$Z_3 = \frac{u_1(l)}{i_1(l)} = -j\cot kl\frac{\rho_1 + 2\rho_2 + (2\rho_2 - 3\rho_1)\tan^2 kl}{3 - \tan^2 kl}. \quad (7.20)$$

If we do not use the assumption (7.15) and to consider that $\rho_3 = 60\,\alpha_3$, the electrodynamics wave impedances are equal to

$$W_{11} = W_{33} = \frac{(\rho_1 - \rho_3)(\rho_1^2 + \rho_1\rho_3 - 2\rho_2^2)}{\rho_1^2 - \rho_2^2}, \quad W_{22} = \frac{\rho_1^2 + \rho_1\rho_3 - 2\rho_2^2}{\rho_1 + \rho_3},$$

$$W_{13} = W_{31} = \frac{(\rho_1 - \rho_3)(\rho_1^2 + \rho_1\rho_3 - 2\rho_2^2)}{\rho_1\rho_3 - \rho_2^2}, \quad W_{12} = W_{21} = W_{23} = W_{32} = \frac{\rho_1^2 + \rho_1\rho_3 - 2\rho_2^2}{\rho_2},$$

and expression for Z_3 will take the form

$$Z_3 = -j\cot kl\frac{A + B\tan^2 kl}{C + D\tan^2 kl}, \quad (7.21)$$

where

$$A = 2\left[\frac{1}{W_{11}} - \frac{3}{W_{12}} + \frac{1}{W_{22}} - \frac{1}{W_{13}}\right]^{-1}, \quad B = A(2\rho_{12} - \rho_{11} - \rho_{13})\left(\frac{1}{W_{11}} - \frac{1}{W_{12}}\right) + 2(\rho_{12} - \rho_{11}),$$

$$C = 2 + A\left(\frac{1}{W_{11}} - \frac{1}{W_{12}} - \frac{1}{W_{13}}\right), \quad D = (2\rho_{12} - \rho_{11} - \rho_{13})\left[\frac{2}{W_{13}} + A\left(\frac{1}{W_{11}} - \frac{1}{W_{12}} - \frac{1}{W_{13}}\right)\left(\frac{1}{W_{11}} - \frac{1}{W_{12}}\right)\right].$$

As calculation results show, impedances Z_3, calculated in accordance with (7.20) and (7.21) for real antennas differ from each other by no more than 1–2%.

As is shown by way of example of calculating the input impedance of three-wire load, use of the approximate expressions (7.16) and (7.17) has little effect on the calculation result, if the inequality $a \ll b \ll l$ is true. At the same time, the application of these approximations greatly simplifies the calculation and allows to use it for load of any number of wires. Determinant Δ_N in accordance with (7.5) is written in the form

$$\Delta_N = \begin{vmatrix} \alpha_1 & \alpha_2 & \alpha_2 & \cdots & \alpha_2 \\ \alpha_2 & \alpha_1 & \alpha_2 & \cdots & \alpha_2 \\ \alpha_2 & \alpha_2 & \alpha_1 & \cdots & \alpha_2 \\ \cdots & \cdots & \cdots & \cdots & \cdots \\ \alpha_2 & \alpha_2 & \alpha_2 & \cdots & \alpha_1 \end{vmatrix}. \quad (7.22)$$

It is easy to check and to see that for N, equal to 1, 2, 3,

$$\Delta_N = (\alpha_1 - \alpha_2)^{N-1}[\alpha_1 + (N-1)\alpha_2]. \tag{7.23}$$

The method of mathematic induction allows to prove that this expression is true for any positive integer N. For this it is necessary to show that the rightness of this expression for the determinant Δ_N means its rightness for the determinant Δ_{N+1}. We expand the determinant by the elements of the first line:

$$\Delta_{N+1} = \alpha_1 \Delta_N + \alpha_2 \sum_{r=2}^{N+1} (-1)^{r+1} M_{1r} = \alpha_1 \Delta_N - N\alpha_2 D_N. \tag{7.24}$$

Here M_{1r} is minor of the determinant Δ_{N+1}, $D_N = M_{12}$ is the determinant of the N-th order:

$$D_N = \begin{vmatrix} \alpha_2 & \alpha_2 & \alpha_2 & \cdots & \alpha_2 \\ \alpha_2 & \alpha_1 & \alpha_2 & \cdots & \alpha_2 \\ \alpha_2 & \alpha_2 & \alpha_1 & \cdots & \alpha_2 \\ \cdots & \cdots & \cdots & \cdots & \cdots \\ \alpha_2 & \alpha_2 & \alpha_2 & \cdots & \alpha_1 \end{vmatrix}.$$

For it the expression similar to expression (7.24) is true:

$$D_N = \alpha_2 \Delta_{N-1} - (N-1)\alpha_2 D_{N-1}.$$

At the same time in accord with (7.22)

$$\Delta_N = \alpha_1 \Delta_{N-1} - (N-1)\alpha_2 D_{N-1}.$$

From the last two equalities

$$D_N = \Delta_N + (\alpha_2 - \alpha_1)\Delta_{N-1}. \tag{7.25}$$

Substituting into (7.25) the value Δ_N from (7.23) and the value Δ_{N-1}, which in accord with (7.23) is equal to $\Delta_{N-1} = (\alpha_1 - \alpha_2)^{N-2}[\alpha_1 + (N-2)\alpha_2]$, we obtained

$$D_N = \alpha_2 (\alpha_1 - \alpha_2)^{N-1}. \tag{7.26}$$

From here in accord with (7.24)

$$\Delta_{N+1} = \alpha_1(\alpha_1 - \alpha_2)^{N-1}[\alpha_1 + (N-1)\alpha_2] - Np_2^2(\alpha_1 - \alpha_2)^{N-1} = (\alpha_1 - \alpha_2)^N(\alpha_1 + N\alpha_2), \tag{7.27}$$

as required.

Since

$$\Delta_{ns} = \begin{cases} \Delta_{n-1}, n = s, \\ (-1)^{n+s}\alpha_2(\alpha_1 - \alpha_2)^{N-2}, n \neq s, \end{cases} \tag{7.28}$$

i.e., electrostatic wave impedances accordingly (7.17) are equal to

$$W_1 = (\rho_1 - \rho_2)\frac{\rho_1 + (N-1)\rho_2}{\rho_1 + (N-2)\rho_2}, \quad W_2 = (\rho_1 - \rho_2)\frac{\rho_1 + (N-1)\rho_2}{\rho_2}. \tag{7.29}$$

Boundary conditions for currents and potentials of N-wire load differ for an even and odd number of wires. For an odd number of wires when $N = 2m - 1$, the boundary conditions are:

$$u_{2n-1}(0) = u_{2n}(0), \quad i_{2n-1}(0) + i_{2n}(0) = 0, \quad i_{2m-1}(0) = 0,$$

$$u_{2n}(l) = u_{2n+1}(l), \quad i_{2n}(l) + i_{2n+1}(l) = 0, \quad u_1(l) = e. \tag{7.30}$$

In this case, as shown in [40], the input impedance of the load is

$$Z_{2m-1} = \frac{u_1(l)}{i_1(l)} = -j(\rho_1 - \rho_2)\tan\beta \frac{2(A_m + A_{m-1} - 1) + H_{2m-1}\left[1 - A_m - A_{m-1} + (B_m + B_{m-1})/(2M\sin^2\beta)\right]}{2(A_{m-1} - A_m) + H_{2m-1}\left[A_m - A_{m-1} + (B_{m-1} - B_m)/(2M\sin^2\beta)\right]}, \tag{7.31}$$

where

$$\beta = kl, \quad A_m = m + 2\sum_{r=1}^{m-1} r\cos 2(m-r)\beta, \quad B_m = \cos 2m\beta,$$

$$M = \frac{\rho_2}{\rho_1 + 2(m-1)\rho_2}, H_{2m-1} = (1 - 4M\sin^2\beta)\bigg/\sum_{s=0}^{m-1}(B_s - 2M_s A\sin^2\beta).$$

Similarly, for an even number of wires ($N = 2m$)

$$Z_{2m} = -j(\rho_1 - \rho_2)\tan\beta\frac{4M\tan^2\beta(C_m + C_{m-1} - \cos^2\beta) + H_{2m}(D_m + D_{m-1})}{4M\tan^2\beta(C_{m-1} - C_m) + H_{2m}(D_{m-1} - D_m)}, \tag{7.32}$$

where

$$C_m = \sum_{r=0}^{m-1}(2r+1)\cos 2(m-r)\beta, D_m \frac{\cos 2(m+1)\beta}{\cos\beta}, H_{2m} = \left[1 - 4M\tan^2\beta\sum_{s=0}^{m-1}C_s\right]\bigg/\sum_{s=0}^{m-1}D_s.$$

For the sake of convenience of calculating, numerators and denominators of expressions (7.31) and (7.32) can be presented as an expansion in powers of the quantity $\tan^2\beta$ [40].

In Table 7.1 expressions for calculating the load impedances with different numbers of wires are given. If $\tan \beta < 1$, the calculation accuracy in accordance with (7.32) and with expressions, presented in Table 7.1, is determined by the number of calculated terms. If $\beta \ll 1$, then, limited in (7.32) by the first two terms of the numerator and denominator, we find:

$$Z_N = -j\frac{\rho_1 + (N-1)\rho_2}{N\beta}\left\{1 - \frac{N\beta^2[N\rho_1 - (N-1)\rho_2]}{3[\rho_1 + (N-1)\rho_2]}\right\}. \tag{7.33}$$

This expression is true and for the odd number of wires. Knowing the load impedance, one can determine all electrical characteristics of the antenna. Figure 7.5a shows the calculated curves for the same antenna with different distance between wires. Dimensions of the antenna (in meters): $L = 200$, $l = 300$, $a = 0.1$. The number of wires in the load is equal to four. The

Table 7.1: Impedances of loads.

Number of wires	Expressions for calculating
2	$-0.5\,j\cot\beta\,[\rho_1 + \rho_2 + (\rho_2 - \rho_1)\tan^2\beta]$
3	$-j\dfrac{\cot\beta}{3-\tan^2\beta}\,[\rho_1 + 2\rho_2 + (2\rho_2 - 3\rho_1)\tan^2\beta]$
4	$-j\dfrac{\cot\beta}{4(1-\tan^2\beta)}\,[\rho_1 + 3\rho_2 - 2(3\rho_1 - \rho_2)\tan^2\beta + (\rho_1 - \rho_2)\tan^4\beta]$
5	$-j\dfrac{\cot\beta}{5-10\tan^2\beta + \tan^4\beta}\,[\rho_1 + 4\rho_2 - 10\,\rho_1\tan^2\beta + (5\rho_1 - 4\rho_2)\tan^4\beta]$
6	$-j\dfrac{\cot\beta}{2(1-3\tan^2\beta)(3-\tan^2\beta)}\times$ $\times\,[\rho_1 + 5\rho_2 - 5\,(3\rho_1 + \rho_2)\tan^2\beta + 3(5\rho_1 - 3\rho_2)\tan^4\beta - (\rho_1 - \rho_2)\tan^6\beta]$

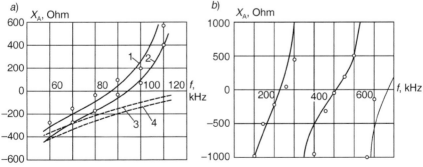

Fig. 7.5: Reactive input impedance of antennas with meandering load of four (a) and six (b) wires.

distance b between the wires axes of the first antenna load is equal to 5, of the second load—to 2. In Fig. 7.5b the results of calculation and experimental verification for an antenna with a meandering load of six wires are shown. Dimensions of the antenna: $L = 50$, $l = 70$, $b = 1.54$, $a = 0.004$. And here the coincidence with the experiment is good. The experiment was performed on the model. The coincidence of experimental values with calculation is good. For comparison, the input reactance of the inverted L antennas with the same sizes is given in the figure by dotted lines (curves 3 and 4). As can be seen from the figures, the use of meandering load increases the electrical length of the antenna and shifts its resonances in the direction of low frequencies.

Figure 7.6a presents the calculated curves and experimental values of the input impedance for similar antenna (curve 1) with six wires in the load and with dimensions: $L = 50$, $l = 45$, $b = 1.54$, $a = 2.5.10^{-2}$. For comparison, the input impedance of the inverted L antennas with the same height and with the length of the horizontal load 90 and 45 m (curves 2 and 3) are given. In Fig. 7.6b calculated efficiency of antennas is given. As seen from the figure, in the range of 200–500 kHz efficiency of the antenna with meandering load is substantially higher than the efficiency of the inverted L antenna of the same size and is comparable with the efficiency of such antenna having double length of the load.

The use of antenna with meandering load in MF range allows to reduce a size of a plot, which the antenna occupies, an area of the grounding and a price of construction.

The first specimen of this antenna was put into operation in 1983 in the town Pavlovo on the radio center of the Baltic Shipping Company. In 1987, the radio center of commercial sea port in Ventspils (Latvia) was equipped with three antennas. The effectiveness of the new antenna has been tested by measuring the field's magnitude, produced by new antennas in comparison with the inverted L antenna. Antennas were operating with

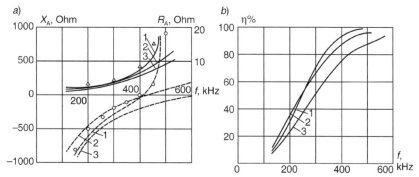

Fig. 7.6: Input impedance (a) and efficiency (b) of antenna with meandering load and inverted L antennas with different length of load.

the same transmitter. Tests have shown that the field created by the antenna with a meandering load is greater in 1.5–1.6 times. This is equivalent to increasing the transmitter power in 2.3–2.6 times.

7.3 Voltages and currents in meandering loads

The correct selection of elements of antenna construction requires calculation of currents and voltages arising therein. For example, the magnitudes of the voltages between the wires ends and the ground determine the choice of insulators, on which the load is suspended. Diameters of wires depend on the maximum currents, etc.

We examine in particular the load consisting of of three wires using condition (7.15) for simplifying calculation. As shown former, substituting the boundary conditions (7.12) in the equations (7.3) for the currents and potentials along the load wires, we obtain equalities (7.19). By means of (7.19) and (7.28) in accordance with the first equation of the system (7.3) one can obtain expressions for the currents along the load wires. The current in a first wire is equal to

$$i_1(z) = I \sin(2\beta + kz), \qquad (7.34)$$

where

$$I = j \frac{U_1}{\cos^2 \beta(\rho_1 + 2\rho_2 - \rho_1 \tan^2 \beta)} = j \frac{e}{\cos^3 \beta[\rho_1 + 2\rho_2 - (3\rho_1 - 2\rho_2)\tan^2 \beta]}.$$

Similarly

$$i_2(z) = -I \sin (2\beta - kz), \ i_3(z) = I \sin kz. \qquad (7.35)$$

The total current in the load wires is

$$i(z) = \sum_{n=1}^{3} i_n(z) = I (2 \cos 2\beta + 1) \sin kz, \qquad (7.36)$$

i.e., current is distributed along coordinate z in accordance with sinusoidal law.

Figure 7.7 shows the distribution of a current along the wires of three-wire load, constructed in accordance with (7.34) and (7.35), and of the total current, constructed in accordance with (7.36). If to straighten mentally the load wire, it will be seen that the current of the straightened wire and the total current are distributed in accordance with sinusoidal law.

Potential does not submit to this law. Expressions (7.3), (7.19) and (7.28) give for the potential another result:

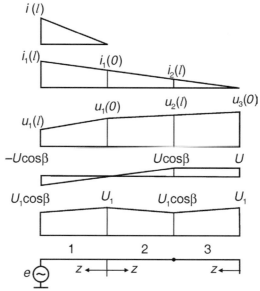

Fig. 7.7: Currents and potentials along the wires of the load.

$$u_1(z) = U_1 \cos kz + j(\rho_1 - \rho_2)I_1 \sin kz = U_1 \cos kz - U \cot \alpha \sin kz, \quad (7.37)$$

where

$$U = \frac{2e(\rho_1 - \rho_2)\tan^2 \beta}{\cos \beta[\rho_1 + 2\rho_2 - (3\rho_1 - 2\rho_2)\tan^2 \beta]}.$$

Similarly

$$u_2(z) = U_1 \cos kz + U \cot \beta \sin kz, \, u_3(z) = U_1 \cos kz + U \cos kz. \quad (7.38)$$

Thus, the potential of the load wire is a sum of two terms. Both terms and its sum are shown in Fig. 7.7.

Sinusoidal law of the current distribution along the load wire remains valid for a greater number of wires. For example, if a load consists of the four wires,

$$i_1(z) = I \sin(3\beta + kz), \, i_2(z) = -I \sin(3\beta - kz), \, i_3(z) = I \sin(\beta + kz),$$
$$i_4(z) = -I \sin(\beta - kz),$$

where

$$I = j\frac{U_1}{\cos^3 \beta[\rho_1 + 3\rho_2 - (3\rho_1 + \rho_2)\tan^2 \beta]} = j\frac{e}{\cos^4 \beta[\rho_1 + 3\rho_2 - 2(3\rho_1 - \rho_2)\tan^2 \beta + (\rho_1 - \rho_2)\tan^4 \beta]}.$$

Table 7.2: Input current of the load.

Number of wires	i_{max}
1	$j\dfrac{e\sin\beta}{\rho_1\cos\beta}$
2	$j\dfrac{e\sin 2\beta}{\cos^2\beta\left[\rho_1+\rho_2-(\rho_1-\rho_2)\tan^2\beta\right]}$
3	$j\dfrac{e\sin 3\beta}{\cos^3\beta\left[\rho_1+2\rho_2-(3\rho_1-2\rho_2)\tan^2\beta\right]}$
4	$j\dfrac{e\sin 4\beta}{\cos^4\beta\left[\rho_1+3\rho_2-2(3\rho_1-\rho_2)\tan^2\beta+(\rho_1-\rho_2)\tan^4\beta\right]}$
5	$j\dfrac{e\sin 5\beta}{\cos^5\beta\left[\rho_1+4\rho_2-10\rho_1\tan^2\beta+(5\rho_1-4\rho_2)\tan^4\beta\right]}$
6	$j\dfrac{e\sin 6\beta}{\cos^6\beta\left[\rho_1+5\rho_2-5(3\rho_1+\rho_2)\tan^2\beta+3(5\rho_1-3\rho_2)\tan^4\beta-(\rho_1-\rho_2)\tan^6\beta\right]}$

Table 7.3: Potential on the end of load wire.

Number of wires	u_{max}
1	$u_1(0)=\dfrac{e}{\cos\chi}$
2	$u_2(l)=\dfrac{e\left[\rho_1+\rho_2+(\rho_1-\rho_2)\tan^2\beta\right]}{\left[\rho_1+\rho_2-(\rho_1-\rho_2)\tan^2\beta\right]}$
3	$u_3(0)=\dfrac{e\left[\rho_1+2\rho_2+(\rho_1-2\rho_2)\tan^2\beta\right]}{\cos\beta\left[\rho_1+2\rho_2-(3\rho_1-2\rho_2)\tan^2\beta\right]}$
4	$u_4(l)=\dfrac{e\left[\rho_1+3\rho_2+2(\rho_1-3\rho_2)\tan^2\beta+(\rho_1-\rho_2)\tan^4\beta\right]}{\left[\rho_1+3\rho_2-2(3\rho_1-2\rho_2)\tan^2\beta+(\rho_1-\rho_2)\tan^4\beta\right]}$
5	$u_5(0)=\dfrac{e\left[\rho_1+4\rho_2+2(\rho_1-6\rho_2)\tan^2\beta+\rho_1\tan^4\beta\right]}{\cos\beta\left[\rho_1+4\rho_2-10\rho_1\tan^2\beta+(5\rho_1-4\rho_2)\tan^4\beta\right]}$
6	$u_6(l)=\dfrac{e\left[\rho_1+5\rho_2+(3\rho_1-23\rho_2)\tan^2\beta+3(\rho_1+\rho_2)\tan^4\beta+(\rho_1-\rho_2)\tan^6\beta\right]}{\left[\rho_1+5\rho_2-5(3\rho_1+\rho_2)\tan^2\beta+3(5\rho_1-3\rho_2)\tan^4\beta-(\rho_1-\rho_2)\tan^6\beta\right]}$

The results of calculating the input current of the load (for different number N of wires) and the potential at the end of the load are given in the Tables 7.2 and 7.3. The presented magnitudes of the current and the potential are maximal in all frequency range up to the series resonance. Therefore, in accordance with these magnitudes it is need to choose the diameters of the wires and the type of insulators.

Figure 7.8 gives as an example the calculated values of the maximum potentials and currents of the load for an antenna with the dimensions: $L = 50$, $l = 45$, $b = 1.54$, $a = 2.5 \cdot 10^{-2}$. The number of wires in the load is $N = 6$. The maximum potential is specified with respect to the voltage e at the load input and to emf at the antenna input. The maximum current is specified with respect to e and to the input current J_A of the antenna. It is taken into account that

$$e_A = e \frac{\cos k(L + l_e)}{\cos kl_e}, J_A = i_1(l)\frac{\sin k(L + l_e)}{\sin kl_e}, \qquad (7.39)$$

where $l_e = \dfrac{1}{k}\tan^{-1}(-jW/Z_N)$ is an equivalent length of the load with input impedance Z_N.

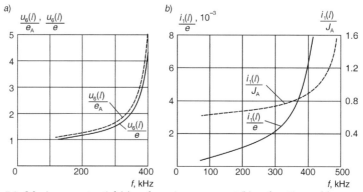

Fig. 7.8: Maximum potential (a) and maximum current (b) as functions of frequency.

8

Folded Antennas, Perpendicular to Metal Surface

8.1 Symmetrical structure

Before proceeding to a main content of the new chapter, it is necessary to make a few remarks. As already mentioned, the creation of the theory of coupled lines was caused by the need to calculate the structures of wires parallel to each other and the infinite metal surface. But with time it was shown that this method allows to find the current distribution along the parallel wires, which are perpendicular to this surface, taking into account the mirror image in this conductive surface, i.e., this method allowed to determine the electrical characteristics of multi-wire vertical structures.

Summing up, it is important to note the general principle underlying the theory of related lines. The theory gives an opportunity to find, in the first approximation, the current distribution along the wire of the linear radiator, which is identical in the first approximation to the current distribution along a uniform long line. The obtained current distribution along the metallic radiating structure coincides with solution of the integral equation for the current in the metal antenna. Later this distribution is used for calculating the active component of the input impedance and for defining more precisely the reactive component of this impedance with the help of the method of the induced emf.

The same approach is used in calculating of impedance antenna, i.e., of radiator with nonzero boundary conditions on its surface: the laws of a current distribution along the impedance radiator and along the equivalent impedance line are also identical. The advantage of using equivalent line is the maximal simplicity of its research and the high efficiency of applying the obtained results to designing radiators with required characteristics.

The theory of related lines is a base for the analysis of antennas, consisting of parallel wires. In particular, the vertical long line of two wires is an equivalent of a folded antenna. Such antenna is a symmetric or asymmetric radiator with parallel wires, spaced on a distance, which is small in comparison with a length of both a wave and a wire. The simplest version of symmetric antenna from two thin wires of an identical diameter, excited in the middle of one of the wires is given in Fig. 8.1. The second (unexcited) wire is continuous (a) or has a gap (b).

Several asymmetrical variants of a folded radiator, situated perpendicular to ground, are shown in Fig. 8.2. These antennas are widespread side by side with conventional linear radiators. In the first such variant the second (unexcited) wire is shorted to the ground (Fig. 8.2a), in the second variant there is a gap between this wire and the ground (Fig. 8.2b). In the general case the diameters of wires forming the radiator are not equal to each other (Fig. 8.2c). The upper ends of the wires of an asymmetric radiator may be connected or not connected and may be positioned at different height (Fig. 8.2d). The radiator may have a shape of coaxial structure (Fig. 8.2e).

An advantage of the folded radiator is in the first place the fact that they are substantially shorter than the linear antennas intended for operating at the same frequency. Selecting variant and dimensions of such antenna provides additional degrees of freedom for obtaining the desired electrical characteristics, for example, for improving matching with signal source (with generator or cable). To sum up, we can conclude that the folded radiators combine the functions of radiation and matching.

The structure of currents in the folded radiator is shown in Fig. 8.3. The conductive metal surface (ground), on which an antenna is installed, causes appearance of displacement currents between the wires and ground (they are shown by dotted lines in Fig. 8.3). The displacement currents create the radiation and cause the emergence of the summand in the input impedance of the antenna, which is analogous to the monopole resistance. They decrease the conduction currents in parallel wires with increasing distance from the ground. Further they are transformed to the conduction currents flowing to the generator pole along the ground surface.

Sometimes engineers name folded antennas as loop antennas. This is mistake. As follows from what has been said the folded antenna has high efficiency whereas the loop antenna is used mainly at reception.

The method of calculating folded antennas is considered using for example an asymmetric folded radiator with a gap in an unexcited wire (see Fig. 8.2b). At first in this gap between the free end of the unexcited wire and the ground we place in parallel two generators of a current with equal magnitude (mJ) and opposite sign (Fig. 8.4). Also, we divide a main

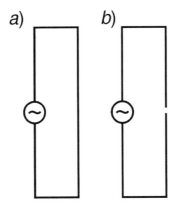

Fig. 8.1: Symmetric folded radiators with continuous unexcited wire (a) and a gap (b).

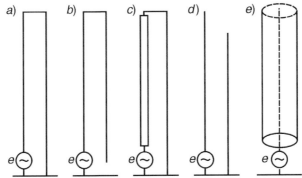

Fig. 8.2: Asymmetric folded radiators: a—with grounding the second wire, b—with a gap in the second wire, c—with wires of different diameter, d—with wires of different length, e—in form of a coaxial structure.

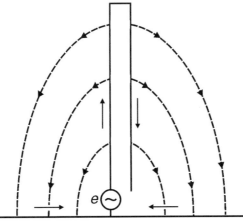

Fig. 8.3: Displacement currents and conduction currents along wires and in the ground near the folded radiator.

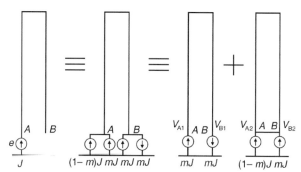

Fig. 8.4: To a calculation of the folded radiator with a gap.

generator (with a current J) on to two parallel generators of identical sign and different magnitude: mJ and $(1 - m)J$.

The voltage at any point of an antenna, for example at point A, in accordance with the principle of superposition is equal to a sum of voltages created in it by all generators. Therefore, as it is shown in Fig. 8.4, the circuit of the folded radiator may be split into two circuits, with two generators in each circuit. The voltages at point A, created in each of these circuits, are calculated and summed. Let the currents of the wires of the first circuit have opposite phases, i.e., phases equal in magnitude and opposite in direction. This means that the first circuit is a two-wire long line shorted at the end. Let the currents of the wires of the second circuit have the same phases, i.e., the potentials of the points of both wires located at the same height (including the points near antenna base) are identical. This means that the second circuit is the linear radiator (monopole), excited at the base.

As follows from Section 7.1, an asymmetric two-wire line located above the ground, which is depicted in Fig. 7.1, is equivalent to a horizontal folded radiator. Currents and potentials in line wires are defined by the system of equations (7.2). Electrostatic and electrodynamics wave impedances included in these equations are defined by equalities (7.4). From the first equation of (7.2) follows that in considered long line there are only anti-phase currents $(i_1 = -i_2)$, if the following conditions are met:

$$I_1 = -I_2, \quad U_1/W_{11} - U_2/W_{12} = -(U_2/W_{22} - U_1/W_{12}),$$

whence in accordance with (7.7) $U_1 = -U_2/g$. From (7.7) and the second equation of (7.2) it follows that the ratio of potentials in the points located on different wires in the same section is equal to $u_1/u_2 = U_1/U_2 = -1/g$. This means that the voltages in points A and B of short-circuited two-wire line are related by the equality

$$V_{A1}/V_{B1} = -1/g. \tag{8.1}$$

If only in-phase currents exist in an asymmetric line, the potentials of wires in the same cross-section of the line should be equal along the entire length of the line ($u_1 = u_2$), i.e.,

$$U_1 = U_2, \quad \rho_{11}I_1 + \rho_{12}I_2 = \rho_{12}I_1 + \rho_{22}I_2,$$

whence $I_1 = gI_2$. This means that the currents of wires in each cross-section of the monopole (including the base of the monopole) are related by

$$i_1/i_2 = J_{A2}/J_{B2} = g. \tag{8.2}$$

From (8.1) and (8.2), we obtain

$$(V - V_{A1})/V_{A1} = (J - J_{B2})/J_{B2} = g,$$

where $V = V_{A1} - V_{B1}$ is the voltage at the input of the line, and $J = J_{A2} + J_{B2}$ is the current at the base of the folded radiator, which is equal to the total current (to the current of the generator). From here

$$V_{A1}/V = J_{B2}/J = m = 1/(1 + g). \tag{8.3}$$

From (8.3) and Fig. 8.4 it is clear that m is a fraction of the in-phase current in the right wire of the monopole.

In the first circuit (in the short-circuited two-wire line) the voltage at the point A is

$$V_{A1} = mV = m^2 J Z_1,$$

where $Z_1 = jW_1 \tan kL$ is the input impedance of the line, W_1 is its wave impedance. In the second circuit (monopole)

$$V_{A2} = J Z_e(a_e),$$

where $Z_e(a_e)$ is the input impedance of the asymmetric linear radiator of height L with an equivalent radius a_e. Dividing the total voltage at the point A by the current of the generator, we find the input impedance of the radiator:

$$Z_A = \frac{V_{A1} + V_{A2}}{J} = Z_e(a_e) + jm^2 W_1 \tan kL. \tag{8.4}$$

As it follows from this expression, the input impedance the folded radiator with a gap is equal to a sum of impedances of a monopole and a short-circuited two-wire long line. These impedances are included in series.

Similarly, the input impedance of the folded radiator with the ground connection of the second wire is a parallel connection of the monopole and the short-circuited long line:

$$Y_A = \frac{1}{jW_1 \tan kL} + \frac{p^2}{Z_e(a_e)}, \qquad (8.5)$$

where $p = 1 - m = g/(1 + g)$ is the fraction of the in-phase current in a left wire of the radiator. In order to obtain (8.5), one must include in a right wire of a folded radiator two generators of voltage, which are disposed in series, equal in magnitude (pe), and opposite in sign. Also, one must divide the main generator (with electromotive force e) on two disposed in series generators of voltage the same in sign, with electromotive forces pe and $(1 - p)e$. Further the circuit of the folded radiator is divided into two circuits—the short-circuited two-wire line and the monopole, and the currents in the base of the left wire created in each of these circuits are calculated and summed.

Expression (8.5) was obtained in [41], where procedure of calculating the folded radiator with wires of different diameters and the ground connection of the second wire is given. If the folded radiator is formed by two identical thin wires with radii $a = a_1 = a_2 \ll b$ (here b is the distance between wires axes—see Fig. 7.3), then

$$m = p = 0.5, \quad g = 1, \ a_e = \sqrt{ab}, \ W_1 = 120 \ln(b/a). \qquad (8.6)$$

For wires with different radii we have

$$p = 1 - m = \frac{\ln(b/a_2)}{2\ln(b/\sqrt{a_1 a_2})}, a_e = \exp \frac{\ln a_1 \ln a_2 - \ln^2 b}{2\ln(\sqrt{a_1 a_2}/b)}, \quad a = \frac{\ln(b/a_2)}{\ln(b/a_1)}, W_1 = 120 \ln \frac{b}{\sqrt{a_1 a_2}}. \qquad (8.7)$$

In the general case

$$m = C_{22}/(C_{11} + C_{22}), \qquad (8.8)$$

where C_{22} is the self capacitance of the right wire, and C_{11} is the self capacitance of the left wire. From (8.3) and (8.7) it follows that the currents in the wires of the monopole are proportional to self capacitances of the wires, and potentials of the wires of the long line are proportional to reactances $X_n = -1/(\omega C_{nn})$ between the wires and the ground.

The limiting case ($m = 1$, $C_{11} = 0$) is shown in Fig. 8.2e. Here folded radiator is designed as a section of a coaxial line, which is open at the bottom and closed at the top. According to (8.4)

$$Z_A = Z_e(a_e) + jW_1 \tan kL,$$

where $W_1 = 60\ln(a_2/a_1)$. If the outer conductor of a coaxial line is shorted to ground, then according to (8.5)

$$Y_A = -j/(W_1 \tan kL).$$

Thus, in extreme cases expressions (8.4) and (8.5) give a sufficiently obvious result.

Input impedance of a folded radiator with the ground connection of the second wire in contrast to the folded radiator with a gap is a parallel connection of a monopole and a short-circuit long line, i.e., it has a more complex character. However, this option allows transforming an active component of a monopole input impedance. Indeed, according to (8.5) the input admittance of the folded radiator at the resonant frequency of a monopole is

$$Y_A = \frac{R_e(a_e) + jp^2 W_l \tan kl}{jR_e(a_e)W_l \tan kL},$$

i.e., neglecting the first term of the numerator as against the second term and substituting the value p in accordance with (8.7), we obtain

$$R_A \approx \frac{R_e(a_e)}{p^2} = R_m(a_e)\left[1 + \frac{\ln(b/a_1)}{\ln(b/a_2)}\right]^2. \tag{8.10}$$

Selecting the radii of the wires permits increasing a level of matching such antenna with a cable or a generator.

Folded antennas have significant advantages over linear radiators: they allow twice to reduce the height, to increase the matching level and to increase the number of resonance frequencies within a given range, not less than in two times. The active and reactive components of the input impedance for symmetrical antennas of this type are twice as large as those

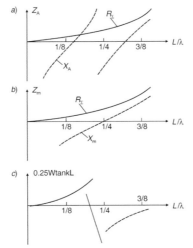

Fig. 8.5: Input impedances of the folded radiator with a gap (a), of the monopole (b) and of the long line (c).

of asymmetric antennas, and the gain is half, that completely coincides with linear antennas.

8.2 Structures with wires of different length and diameter

The previous sections are devoted to parallel long lines and to folded radiators consisting of wires with the equal length. Procedure of calculating a current distribution along wires of long line is based on the theory of electrically coupled lines. In order to determine the input impedance of a folded antenna, its circuit is divided into a two-wire long line and a linear radiator. If an asymmetrical folded radiator is formed by wires of different length (see Fig. 8.2d), the structure may be similarly split into the line and the radiator, but the method of calculating current in the line and the radiator becomes not obvious. This is not the sole task that requires determining electrical characteristics of a line and a radiator, consisting of wires with different lengths.

An example of a long line with wires of different lengths is given in Fig. 8.6a. As is seen from the figure, it is distinguished from the long line shown in Fig. 8.2a and consisting of two sections. The lengths of the upper and lower sections are equal respectively to $l = l_1 - l_2$ and $L = l_2$, where l_1 is the length of a longer wire and l_2 is the length of a shorter wire. The lower section has the form of two parallel wires of the same length with the same cross section, for example the form of a circle with a radius a. Capacitance per unit length between two thin wires located in a homogeneous medium with permittivity ε is equal to

$$C_0 = \pi\varepsilon/\ln(b/a), \tag{8.11}$$

where b is the distance between the wires axes. This capacitance determines the wave impedance of the lower section (of two-wire long line).

The presence of the upper section leads to a change in its input impedance and can be determined by means of calculating and measuring the end-of-line capacitance of the lower section. The electrostatic structure in

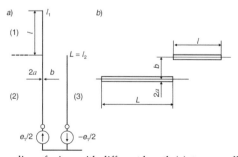

Fig. 8.6: A long line of wires with different length (a); two parallel wires (b).

this case consists of three conducting elements designated as (1), (2) and (3) in Fig. 8.6a. One must calculate capacitance between elements (2) and (3) in the presence of element (1). This capacitance is not equal to the capacitance between isolated elements (2) and (3) in the absence of an element (1). For this reason, one must find the necessary capacitance as the difference of two capacitances:

$$C = C_1 - C_0 L, \tag{8.12}$$

where C_1 is the total capacitance between the long and short wire, and $C_0 L$ is the capacitance between the wires of the lower section. At that C_1 is the capacitance of an electrically neutral system consisting of two conductors (see, for example, [24]):

$$C_1 = (\alpha_{11} + \alpha_{22} - 2\alpha_{12})^{-1}, \tag{8.13}$$

where α_{ik} are potential coefficients, calculated by the following formulas:

$$\alpha_{11} = \frac{1}{2\pi\varepsilon L}\left\{\ln\left[L/a + \sqrt{1+(L/a)^2}\right] + a/L - \sqrt{1+(a/L)^2}\right\},$$

$$\alpha_{22} = \frac{1}{2\pi\varepsilon(L+l)}\left\{\ln\left[(L+l)/a + \sqrt{(L+l)^2/a^2 + 1}\right] + a/(L+l) - \sqrt{a^2/(L+l)^2 + 1}\right\},$$

$$\alpha_{12} = \frac{1}{4\pi\varepsilon(L+l)}\left\{\ln\left[\left(L + \sqrt{L^2 + b^2}\right)/b\right] + (L+l)\ln\left[\left(L+l+\sqrt{(L+l)^2 + b^2}\right)/b\right]\right\}/L -$$

$$- \sqrt{L^2 + b^2}/L + \sqrt{l^2 + b^2}/L - l\ln\left[\left(l + \sqrt{l^2 + b^2}\right)/b\right]/L + b/L - \sqrt{(L+l)^2 + b^2}/L\}.$$

Since $L/a, l/a \gg 1$, then

$$\alpha_{11} = \frac{1}{2\pi\varepsilon L}\left(\ln\frac{2L}{a} - 1\right), \alpha_{22} = \frac{1}{2\pi\varepsilon(L+l)}\left[\ln\frac{2(L+l)}{a} - 1\right].$$

Sometimes inequalities $L/b, l/b \gg 1$ take place. In this case the expression for α_{12} gets simplified:

$$\alpha_{12} = \frac{1}{4\pi\varepsilon(L+l)}\left[\ln\frac{2L}{b} + \ln\frac{2(L+l)}{b} + \frac{l}{L}\ln\frac{L+l}{l} - 2\right].$$

Calculations show that C is small as compared with $C_0 L$. In particular, if wires are located in the air, i.e., $\varepsilon_0 = 1/(36\pi \cdot 10^9)$, and $L = 7.5$, $b = 1.0$, $2a = 0.05$, when l changes from 1 to 4 (all dimensions are in centimeters), we have $C_0 L = 7.07$ pF, and C changes from 0.05 to 0.1 pF. Thus, the additional length l creates capacitive load on the end of the two-wire line. It is equivalent to lengthening this long line by a section with a length

$$l_0 = (1/k)\cot^{-1}[1/(\omega CW_l)], \qquad (8.14)$$

where W_l is the wave impedance of the long line. The calculation results, showing the values of capacitance C, and also the equivalent section lengths l_0 for the mentioned dimensions at 1 GHz, are given in Table 8.1.

It is easily convinced that the capacitance between the elements (2) and (3) in the absence of the element (1) is substantially greater than the capacitance presented in Table 8.1.

These calculations were verified by simulations with the help of CST program. The model of structure, which was applied at this simulation, is shown in Fig. 8.7, where e is a generator with output impedance $R = 50$ Ohm. The simulation results for the lengthened section, denoted as l_{01}, are also presented in Table 8.1. Since the distance b between the wires is finite, then the dimensions l_0 and l_{01} for $l = 0$, obtained on the basis of approximate theory of two-wire long line, differ from 0. A cause of this fact are self capacitances of the wires. In order to clearly demonstrate the magnitudes l and an effect on lengthening section, dimensions l_0 and l_{01} are decreased by their values at $l = 0$. As it is seen from Table 8.1, the calculation and simulation results agree well for $1 \leq 0.1\lambda$. Results show that the input impedance of a line with wires of unequal lengths differs comparatively

Table 8.1: Capacitive load due to the unequal wires.

l, cm	$2a = 0.05$ cm			$2a = 0.2$ cm		
	l_0, cm	l_{01}, cm	C, pF	l_0, cm	l_{01}, cm	C, pF
0.0	0	0	0.020	0	0	0.047
0.5	0.22	0.19	0.037	0.21	0.15	0.073
1.0	0.41	0.39	0.050	0.37	0.30	0.093
1.5	0.56	0.52	0.063	0.49	0.45	0.108
2.0	0.69	0.86	0.073	0.58	0.61	0.119
2.5	0.80	1.10	0.081	0.65	0.79	0.128
3.0	0.90	1.38	0.089	0.71	1.00	0.135
3.5	0.98	1.66	0.095	0.75	1.24	0.140
4.0	1.05	1.94	0.101	0.78	1.48	0.144
4.5	1.12	2.17	0.107	0.81	1.64	0.148

Fig. 8.7: The simulation model for the two-wire long line with wires of different length.

weakly from the input impedance of a two-wire line, whose length is equal to the length of the shorter wire.

Similar results at $2a = 0.2$ cm are given also in Table 8.1.

In accordance with obtained results one can write the current distributions along the wires of line:

$$i_1(z) = \begin{cases} I_0 \sin kl_0 \sin k(L+l-z)/\sin kl, & L \leq z \leq L+l, \\ I_0 \sin k(L+l_0-z), & 0 \leq z \leq L, \end{cases}$$

$$i_2(z) = \begin{cases} 0, & L \leq z \leq L+l, \\ -I_0 \sin k(L+l_0-z), & 0 \leq z \leq L, \end{cases} \qquad (8.15)$$

where I_0 is the generator current. A long line with equal length of wires located in free space radiates only in the case when the distance between the wires is not too small compared with the wave length. In case of wires of unequal length, the additional segment l of the longer wire radiates a signal, as it follows from expressions (8.15).

The obtained results allow to consider another problem—calculating input impedance of a linear radiator (monopole) composed of two parallel wires with different lengths (Fig. 8.8a).

In this case it is necessary to divide the equivalent line into two sections, as shown in Fig. 8.8b. The section 1 has one wire; the section 2 consists of two wires. The section number is indicated in parentheses, the wire number is indicated at its base. The currents and potentials along a section m of a wire n in the asymmetric line are given by (7.8), where $n = 1, 2, m = 1, 2$. If the distance between the wires is small in comparison with the wires lengths, one can consider that

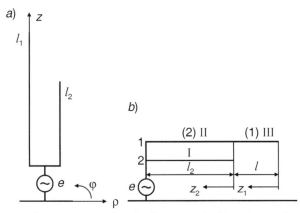

Fig. 8.8: Monopole, formed by the wires of different lengths (a), and equivalent long line (b).

$$\rho_{nn}^{(m)} = const(n) = \rho_1^{(m)}, \ \rho_{ns}^{(m)}\Big|_{n \neq s} = const(n) = \rho_2^{(m)},$$
$$W_{nn}^{(m)} = const(n) = W_1^{(m)}, \qquad W_{ns}^{(m)}\Big|_{n \neq s} = W_2^{(m)}.$$

The zero currents at the ends of the wires and the continuity of the current and potential along each wire permit to write the boundary conditions:

$$i_1^{(1)}\Big|_{z_1=0} = i_2^{(2)}\Big|_{z_3=0} = 0; \ i_1^{(1)}\Big|_{z_1=l} = i_1^{(2)}\Big|_{z_2-0};$$
$$u_1^{(1)}\Big|_{z_1=l} = u_1^{(2)}\Big|_{z_2=0}; u_1^{(2)}\Big|_{z_2=l_2} = u_2^{(2)}\Big|_{z_2=l_2} = e.$$

From these boundary conditions we get

$$I_1^{(1)} = I_2^{(2)} = 0; I_1^{(2)} = j\frac{U_1^{(1)}}{W_1^{(1)}}\sin kl; U_1^{(2)} = U_1^{(1)}\cos kl;$$

$$U_2^{(2)} = U_1^{(1)}\left[\cos kl - \frac{\rho_1^{(2)} - \rho_2^{(2)}}{W_1^{(1)}}\sin kl \cdot \tan kl_2\right];$$

$$U_1^{(1)} = \frac{e}{\cos kl \cos kl_2}\left[1 - \frac{\rho_1^{(2)}}{W_1^{(1)}}\tan kl \tan kl_2\right]^{-1}.$$

The current distribution along the first section of the longer wire is given by

$$i_1^{(1)} = j\frac{U_1^{(1)}}{W_1^{(1)}}\sin k(l_1 - z), \tag{8.16}$$

The current along the second section is

$$i_1^{(2)} = jU_1^{(1)}\left\{\frac{\sin kl \cos kz_2}{W_1^{(1)}} + \cos kl\left[\frac{1}{W_1^{(2)}} - \frac{1}{W_2^{(2)}}\left(1 - \frac{\rho_1^{(2)} - \rho_2^{(2)}}{W_1^{(1)}}\right)\tan kl \tan kl_2\right]\sin kz_2\right\}. \tag{8.17}$$

And the current along the shorter wire

$$i_2^{(2)} = jU_1^{(1)}\cos kl\left\{\frac{1}{W_1^{(2)}} - \frac{1}{W_2^{(2)}}\left[1 - \frac{\rho_1^{(2)} - \rho_2^{(2)}}{W_1^{(1)}}\right]\tan kl \tan kl_2\right\}\sin kz_2. \tag{8.18}$$

The total current along the second section is

$$i_1^{(2)} + i_2^{(2)} = jU_1^{(1)}\left\{\frac{\sin kl \cos k(l_2 - z)}{W_1^{(1)}} + \cos kl\left[\frac{1}{W_1^{(2)}} - \frac{1}{W_2^{(2)}}\left(1 - \frac{\rho_1^{(2)} - \rho_2^{(2)}}{W_1^{(1)}}\right)\tan kl \tan kl_2\right]\sin k(l_2 - z)\right\}. \tag{8.19}$$

These expressions show that the current distribution along both sections of the monopole is sinusoidal, i.e., similar to the current distribution along a monopole consisting of two sections with different wave impedances (for example, with different wire diameters).

Let us write the expression for the total current along the monopole in the form

$$J_{Am}(z) = \sum_{n=1}^{M} i_n^{(m)}(z), l_{m+1} \le z \le l_m, \tag{8.20}$$

where $i_n^{(m)}(z) = A_{nm} \cos k(l_m - z) + jB_{nm} \sin k(l_m - z)$. In accordance with (8.16)–(8.18)

$$A_{11} = A_{21} = A_{22} = B_{21} = 0, \quad B_{11} = U_1^{(1)}/W_1^{(1)}, \quad A_{12} = jU_1^{(1)} \sin kl/W_1^{(1)},$$

$$B_{12} = B_{22} = U_1^{(1)} \cos kl \left[\frac{1}{W_1^{(2)}} - \frac{1}{W_2^{(2)}} \left(1 - \frac{\rho_1^{(2)} - \rho_2^{(2)}}{W_1^{(1)}} \right) \tan kl \tan kl_2 \right]. \tag{8.21}$$

The input reactance of the monopole is equal to the input impedance of the equivalent line:

$$jX_A = Z_I = \frac{e}{J_A(0)} = -j \frac{\left[W_1^{(1)} - \rho_1^{(2)} \tan kl \tan kl_2 \right] \cos^2 kl_2}{\tan kl \cos^2 kl_2 + DW_1^{(1)} \sin 2kl_2}, \tag{8.22}$$

where $D = \frac{1}{W_1^{(2)}} - \left[1 - \frac{\rho_1^{(2)} - \rho_2^{(2)}}{W_1^{(1)}} \right] \frac{\tan kl \tan kl_2}{W_2^{(2)}}.$

The radiation resistance of the monopole is equal to

$$R_{\Sigma} = 40 \, k^2 h_e^2. \tag{8.23}$$

where h_e is the monopole's effective height given by

$$h_e = \frac{1}{kJ(0)} \sum_{m=1}^{2} \left\{ (A_{1m} + A_{2m}) \sin k (l_m - l_{m+1}) + j (B_{1m} + B_{2m}) \left[1 - \cos k (l_m - l_{m+1}) \right] \right\} =$$

$$\frac{jU_1^{(1)}}{kJ(0)W_1^{(1)}} \left\{ 1 + \sin kl \sin kl_2 + \cos kl \left[2DW_1^{(1)} (1 - \cos kl_2) - 1 \right] \right\}. \tag{8.24}$$

These expressions define the currents along each wire of asymmetric radiator and allow to determine more accurately its input impedance. Considering that an antenna is linear radiator and the current along it is equal to a total current of both wires, it is possible to find the input impedance, for example, by the method of induced emf (second formulation). During calculating the tangential component of the electric field, one must take into account the discontinuity of the current derivative on the sections boundaries.

8.3 Losses in the ground

Application of folded radiators largely depends on their losses magnitude, particularly losses in the ground. Each of the elements, of which antenna consists (i.e., a monopole and a long line), has losses in a ground. Loss resistance R_{ge} for a monopole is calculated in the usual manner. With regard to a long line, its loss resistance R_{gl} in the ground is also non-zero. There is an area, in which fields, created by currents of two closely spaced parallel wires of the long line, have great amplitudes and opposite direction. A radius of this area is relatively small. However, it is necessary to take into account that the magnetic field increases as a result of approaching to a conductor with a current, and thus a current density in ground, equal in a first approximation to the tangential component of the magnetic field, increases, and losses of electromagnetic energy per unit area increase also.

Let J_1 and J_2 are the currents in left and right wire of the long line:

$$J_1 = -J_2 = mJ \frac{\sin k(l+z)}{\sin kl},$$

where mJ is the current in the base of the wire and $l = \lambda/4 - L$ is the wire length (Fig. 8.9a). Then the vector potential of the electromagnetic field produced by the current J_1 at an arbitrary point on the ground surface with coordinates $(x, y, 0)$ at a distance $\rho = \sqrt{x^2 + y^2}$ from the axis of the first wire in view of mirror image

$$A_{z1} = \frac{mJ\mu}{2\pi \sin kL} \int_0^L \sin k(l+z) \frac{\exp\left(-jk\sqrt{\rho^2 + z^2}\right)}{\sqrt{\rho^2 + z^2}} dz.$$

The tangential components of a magnetic field

$$H_{x1} = \frac{\partial A_{z1}}{\mu \partial y} = \frac{mJ}{2\pi j \sin kL} yA_0(L,l,\rho), \quad H_{y1} = -\frac{\partial A_{z1}}{\mu \partial x} = \frac{mJ}{2\pi j \sin kL} xA_0(L,l,\rho),$$

(8.25)

where

$$A_0(L,l,\rho) = -\frac{1}{\rho^2}\left[jL\frac{\exp(-kR)}{R}\sin k(l+L) + \exp(-jkR)\cos k(l+L) + \exp(-jk\rho)\cos kl \right],$$

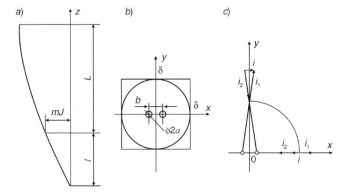

Fig. 8.9: Current distribution along the wire (a), area of losses (b) and currents on a ground surface in the outer area (c).

and $R = \sqrt{\rho^2 + L^2}$. In calculating an integral the substitution $R = \sqrt{\rho^2 + L^2}$ was used.

Let the coordinates' origin coincides with the middle of an interval between the wires. Then the components of a magnetic field strength created by wires of the long line, and the components of a current density in the ground, equal to this strength, are

$$i_x = -H_y = \frac{mJ}{2\pi j \sin kL}\left[(x + b/2)A_0(L,l,R_1) - (x - b/2)A_0(L,l,R_2)\right],$$

$$i_y = -H_x = \frac{mJy}{2\pi j \sin kL}\left[A_0(L,l,R_1) - A_0(L,l,R_2)\right],$$

where $R_{1(2)} = \sqrt{(x \pm b/2)^2 + y^2}$.

Introducing a value δ satisfying the inequality

$$b/2 \ll \delta \ll L, \lambda/2, \tag{8.26}$$

one can divide the area of losses (Fig. 8.9b) into two areas, the boundary between which is a circumference of radius δ. In the outer area the distance from a long line to the observation point is large in comparison with the distance between the wires. Here the fields produced by the currents J_1 and J_2 cancel each other (Fig. 8.9c). Thus, for example, when $x = 0$, $i = -i_1 b/\rho \approx -bH_{x1}/\rho$, and taking into account (8.25), we obtained

$$|i| \le \frac{1,5mJ}{\pi \sin kL}\frac{b}{\rho^2}.$$

Power of losses in the outer area

$$P_I = R_0 \int_0^{2\pi}\int_0^{\infty} |i|^2 \, \rho d\rho d\varphi \le \pi R_0 \left(\frac{1,5mJb}{\pi\delta \sin kL}\right)^2,$$

where $R_0 = 1/(s\sigma)$ is resistance per unit area of the ground surface. In accordance with (8.26), this value is small and can be neglected. In the inner area $\rho \ll L, \lambda/2$ expressions for the components of the surface current are greatly simplified:

$$i_x = \frac{mJ}{2\pi}\left[\frac{x-b/2}{(x-b/2)^2 + y^2} - \frac{x+b/2}{(x+b/2)^2 + y^2}\right], \quad i_y = \frac{mJy}{2\pi}\left[\frac{1}{(x-b/2)^2 + y^2} - \frac{1}{(x+b/2)^2 + y^2}\right].$$

$$(8.27)$$

Without wasting adopted accuracy, we calculate the losses not in a circle of radius δ, but in a square with side 2δ (see Fig. 8.9b). Power of losses in the inner area is

$$P_{II} = 4R_0 \int_0^{\delta}\int_0^{\delta}(i_x^2 + i_y^2)dx\,dy = \frac{R_0(mJb)^2}{\pi^2}\int_0^{\delta}\int_0^{\delta}\frac{dx\,dy}{[(x-b/2)^2 + y^2][(x+b/2)^2 + y^2]},$$

whence $R_{gl} = \dfrac{P_{II}}{(mJ)^2} = \dfrac{R_0}{\pi^2}\displaystyle\sum_{i=1}^{2}Q_i,$

where $Q_1 = -\displaystyle\int_0^{1}\left[\cot^{-1}(t-b/2\delta) + \cot^{-1}(t+b/2\delta)\right]\frac{dt}{t}, \quad Q_2 = \displaystyle\int_0^{1}\left[\frac{\cot^{-1}(t-b/2\delta)}{t-b/2\delta} + \right.$

$\left.\dfrac{\cot^{-1}(t+b/2\delta)}{t+b/2\delta}\right]dt .$

One can show that

$$\sum_{i=1}^{2}Q_i = \pi \ln(b/2a) + 2G,$$

where $G = \displaystyle\int_0^{1}\tan^{-1}z\frac{dz}{z} \approx 0,916$ is Catalan's constant. The magnitude δ, as one would expect, did not enter into this expression. Taking into account that $b/a \gg 1$, we find

$$R_{gl} = \frac{R_0}{\pi}\ln(b/a) = \frac{11}{\sqrt{\sigma\lambda}}\ln(b/a). \qquad (8.28)$$

This result can also be obtained using known analogy between electric field in a conductive medium and an electrostatic field. The rightness of this approach follows from (8.27): a magnetic field and current density in the ground in the area of the losses are determined only by the currents in the base of wires and have a quasi-static nature.

Similarly, one can obtain for the coaxial line (see Fig. 8.2e)

$$R_{gl} = \frac{R_0}{2\pi} \ln(a_2/a_1) = \frac{5.5}{\sqrt{\sigma\lambda}} \ln(b/a). \tag{8.29}$$

As can be seen from (8.28) and (8.29), the loss resistance in the ground for vertically located long line (both two-wire and coaxial) does not depend on its length L. This resistance depends only on the ratio b/a of the distance b between the wires to the wire radius a. Figure 8.10 demonstrates the dependence of loss resistance R_{gl} of two-wire line on the frequency at different values of the ratio b/a and of a conductivity σ of a sea water. The calculation shows that ignoring the losses in the water is an error.

One must emphasize that the loss resistance R_{gl} of the monopole in the ground should be included in series with the input impedance of the radiator itself, and the loss resistance R_{gl} of the line in the ground should be included in series with the input impedance of the line. This is easily to check by means of Fig. 8.11, where the contour along which the current flows, is shown for both antenna elements. Taking into account losses in the ground the expression for the input impedances of folded radiator with a gap and for the input admittance of folded radiator with shorting to ground take the form:

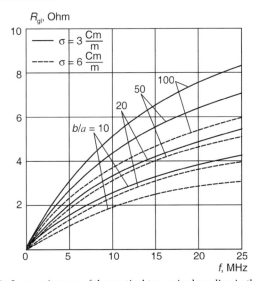

Fig. 8.10: Loss resistance of the vertical two-wire long line in the water.

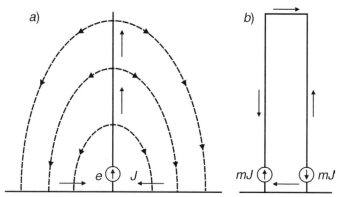

Fig. 8.11: Direction of currents in monopole and ground (a), in two-wire line and ground (b).

$$Z_A = Z_e(a_e) + R_{ge} + jm^2 W_l \tan kL + m^2 R_{gl}, \quad Y_A = \frac{1}{R_{gl} + jW_l \tan kL} + \frac{p^2}{R_{ge} Z_e(a_e)}.$$
(8.30)

Loss resistances of folded radiators in the wires are considered in Section 9.2.

8.4 Impedance folded radiators

Folded radiator, on whose surface in contrast to metal radiator nonzero boundary conditions are performed, is called by impedance folded radiator. Non-zero boundary conditions outside an excitation zone have the form:

$$\left.\frac{E_z(a,z)}{H_\varphi(a,z)}\right|_{-L \le z \le L} = Z(z).$$
(8.31)

Here $E_z(a,z)$ and $H_\varphi(a,z)$ are the tangential component of the electric field and the azimuthal component of the magnetic field respectively, $Z(z)$ is the surface impedance, which in the general case depends on the coordinate z and substantially changes the distribution of current along an antenna in the first approximation. The boundary conditions of this type are true, if the structure of a field in one of a media (for example, inside the magneto dielectric sheath of the antenna) is known and does not depend on a field structure in a different environment (surrounding space). Using impedance structures (or concentrated loads) creates an additional degree of freedom and permits to expand opportunities of the antenna [1]. Two asymmetric versions of the impedance folded radiator are the radiators, in which an unexcited (passive)wire has a gap or is shorting to the ground. They are shown in Fig. 8.12.

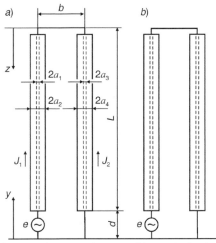

Fig. 8.12: Impedance folded radiators with a second (unexcited) wire, which has a gap (a) or is shorting to the ground (b).

The method of calculating an impedance folded radiator, similar to calculation of the metal folded radiator, is based on the theory of asymmetric lines [42]. For two-wire impedance long line, consisting of wires with impedance coating and located above ground, telegraph equations generalizing equations (7.1), are correct:

$$-\frac{\partial u_1}{\partial z} = j\left(X_{11} + Q_1\right)i_1 + jX_{12}i_2, \qquad u_1 = j\frac{1}{k^2}\left(X_{11}\frac{\partial i_1}{\partial z} + X_{12}\frac{\partial i_2}{\partial z}\right),$$

$$-\frac{\partial u_2}{\partial z} = j\left(X_{22} + Q_2\right)i_2 + jX_{12}i_1, \qquad u_2 = j\frac{1}{k^2}\left(X_{22}\frac{\partial i_2}{\partial z} + X_{12}\frac{\partial i_1}{\partial z}\right).$$

$$(8.32)$$

Here, as in Section 7.1, u_i is the potential of the wire i relative to ground, i_i is the current along the wire i, $X_{ik} = \omega\Lambda_k$ is the self or mutual inductive impedance per unit length. Besides that, $jQ_i = Z_i/(2\pi a_i)$ is additional impedance per unit length created by surface impedance Z_i.

The two left equations of the set (8.32) are based on the fact that the potential decrease at segment dz of each wire is the result of the emf influence. The emfs are induced by the self current and by the currents of adjacent wires. The other two equations are written on the basis of the electrostatic equations relating charges and potentials in accordance with the equation of continuity. The dependence of a current on coordinate z is adopted in the form $\exp(\gamma z)$.

Differentiating the right pair of equations and substituting them into the left equations, we find:

$$\left(X_{11}+Q_1+\frac{\gamma^2}{k^2}X_{11}\right)i_1+X_{12}\left(1+\frac{\gamma^2}{k^2}\right)i_2=0,\ X_{12}\left(1+\frac{\gamma^2}{k^2}\right)i_1+\left(X_{22}+Q_2+\frac{\lambda^2}{k^2}X_{22}\right)i_2=0.$$

$$(8.33)$$

In order to this set of uniform equations has a solution, it is necessary to satisfy the condition

$$\gamma^2_{1(2)}=-k^2-\frac{\omega}{2}\left[G_1\beta_{11}+Q_2\beta_{22}\mp\sqrt{(Q_1\beta_{11}-Q_2\beta_{22})^2+4Q_1Q_2\beta_{12}^2}\right]. \quad (8.34)$$

Here $\beta_{11}=\dfrac{k^2X_{22}}{\omega(X_{11}X_{22}-X_{12}^2)}, \beta_{12}=-\dfrac{k^2X_{12}}{\omega(X_{11}X_{22}-X_{12}^2)}, \beta_{22}=\dfrac{k^2X_{11}}{\omega(X_{11}X_{22}-X_{12}^2)}$

are coefficients of electrostatic induction. Thus, a system of two non-metal wires located above the ground, has two different propagation constants. If the impedance is a purely reactive, they are equal

$$k_{1(2)}=\sqrt{-\gamma^2_{1(2)}}.$$

We seek a solution in the form

$$u_1=A\cos k_1z+jB\sin k_1z+C\cos k_2z+jD\sin k_2z. \quad (8.35)$$

The ratio of the currents obtained from (8.33) is substituted in the first equation of (8.32):

$$i_1=j\frac{k^2+\gamma^2}{\gamma^2Q_1}\frac{du_1}{dz}=a_1(B\cos k_1z+jA\sin k_1z)+a_2(D\cos k_2z+jC\sin k_2z), \quad (8.36)$$

where $a_i=(k^2-k_i^2)/(k_iQ_i)=-1/W_i$. From (8.36) and similar relationship between i_2 and u_2 one can obtain

$$\frac{du_2}{dz}:\frac{du_1}{dz}=\frac{Q_2i_2}{Q_1i_1}, \quad (8.37)$$

i.e.,

$$u_2=b_1(A\cos k_1z+jB\sin k_1z)+b_2(C\ co\ k_2z+jD\sin k_2z), \quad (8.38)$$

and $b_i=\dfrac{Q_2}{Q_1}\dfrac{k^2(X_{11}+Q_1)-k_i^2X_{11}}{(k_i^2-k^2)X_{12}}$. Finally, from (8.37) and (8.36) it follows

that $i_2=a_1c_1(B\cos k_1z+jA\sin k_1z)+a_2c_2(D\cos k_2z+jC\sin k_2z), \quad (8.39)$

where $c_i = b_i Q_1 / Q_2$.

Setting $z = 0$, we find that $A = U_{11}$, $D = I_{11}/a_1$, $C = U_{12}$, $D = I_{12}/a_2$, where U_{11}, U_{12}, I_{11}, I_{12} are fractions of voltages and currents at a beginning of the first wire (near load), which correspond to phase constant k_1 and k_2. Considering that the current flowing from the generator to the load, i.e., in the direction of negative z, is a positive current, we rewrite (8.35), (8.36), (8.38) and (8.39), taking into account the defined coefficients:

$$u_1 = U_{11} \cos k_1 z + U_{12} \cos k_2 z + j\left(W_1 I_{11} \sin k_1 z + W_2 I_{12} \sin k_2 z\right),$$

$$i_1 = I_{11} \cos k_1 z + I_{12} \cos k_2 z + j\left(\frac{U_{11}}{W_1} \sin k_1 z + \frac{U_{12}}{W_2} \sin k_2 z\right),$$

$$u_2 = b_1 U_{11} \cos k_1 z + b_2 U_{12} \cos k_2 z + j\left(b_1 W_1 I_{11} \sin k_1 z + b_2 W_2 I_{12} \sin k_2 z\right),$$

$$i_2 = c_1 I_{11} \cos k_1 z + c_2 I_{12} \cos k_2 z + j\left(c_1 \frac{U_{11}}{W_1} \sin k_1 z + c_2 \frac{U_{12}}{W_2} \sin k_2 z\right).$$

$$(8.40)$$

One must emphasize the important conclusion, which follows directly from (8.40). Currents and potentials of both wires are connected by rigid relations depending not on the details of combining antenna elements into an overall structure (not in accordance with so-called boundary conditions) but depending on the wires diameters and the surface impedance. Therefore, changing the boundary conditions (for example, the point of connecting emf, magnitudes and points of loads placement), it is impossible to create only in-phase or anti-phase currents in the wires of the impedance folded radiator (by contrast to purely metal folded radiators). Accordingly, the input impedance of such impedance radiator cannot be presented as an aggregate of impedance lines and radiators, which connected in parallel or in series. Only folded radiators consisting of identical wires, as it will be shown further, are an exception.

We apply these results to the calculating input impedance of a folded radiator with a gap in second (unexcited) wire (see Fig. 8.12a). The boundary conditions for this variant have the form:

$$i_1(0) + i_2(0) = 0, \; u_1(0) = u_2(0), \; i_2(L) = 0, \; u_1(L) = e. \tag{8.41}$$

Substituting (8.40) into (8.41), we find

$$I_{12} = -I_{11}\frac{1+c_1}{1+c_2}, U_{12} = U_{11}\frac{1-b_1}{b_2-1}, I_{11} = -jd_1\frac{U_{11}}{W_1},$$

$$e = U_{11}\left(\cos k_1 L - \frac{1-b_1}{1-b_2}\cos k_2 L\right) + I_{11}\left(W_1\sin k_1 L - W_2\frac{1+c_1}{1+c_2}\sin k_2 L\right),$$

$$(8.42)$$

and

$$d_1 = \tan k_1 L\left[1 - \frac{W_1(1-b_1)c_2\sin k_2 L}{W_2(1-b_2)c_1\sin k_1 L}\right] : \left[1 - \frac{c_2(1+c_1)\cos k_2 L}{c_1(1+c_2)\cos k_1 L}\right].$$

Then the input impedance of impedance line is

$$X_{il} = \frac{e}{ji_1(L)} = -W_1\cot k_1 L \frac{1 - \frac{(1-b_1)\cos k_2 L}{(1-b_2)\cos k_1 L} + d_1\tan k_1 L\left[1 - \frac{W_2(1+c_1)\sin k_2 L}{W_1(1+c_2\sin kL_1)}\right]}{1 - \frac{W_1(1-b_1)c_2}{W_2(1-b_2)c_1} - d_1\cot k_1 L\left[1 - \frac{(1+c_1)\cos k_2 L}{(1+c_2)\cos k_1 L}\right]}.$$

$$(8.43)$$

Expression (8.43) makes it possible to determine approximately reactive component of the input impedance of the impedance folded radiator (similarly to the fact as formula for the input impedance of an equivalent long line allows to determine a reactive component of an input impedance of the line radiator). The antenna input impedance can be found more precisely by the method of induced electromotive force. Equating the oscillating power passing through a closed surface surrounding the antenna and the oscillating power passing through the source of emf, we obtain (for asymmetric radiator)

$$Z_A = -\frac{1}{J_1^2(0)}\int_0^L E_y J(y)dy,$$

$$(8.44)$$

where E_y is a field on the antenna surface, $J_1(0)$ is a current of a generator and $J(y) = J_1(y) + J_2(y)$ is a total current of an antenna as function of coordinate $y = L - z$ (see Fig. 8.12a).

Expression (8.44) is a generalization of the method of induced electromotive force for the folded radiator. In the folded radiator with a gap a total input current of the antenna coincides with a generator current. In the folded radiator with shorting to a ground this coincidence is absent. In the vicinity of parallel resonance, where the first formulation of the method of induced emf gives the wrong result, for the folded radiator with a gap one must use the expression

$$Z_A = \frac{e}{2J_1(0) + \dfrac{1}{e}\displaystyle\int_0^L E_y(y)J(y)dy}, \qquad (8.45)$$

and the total current of this antenna is equal to

$$J(y) = j(1 + c_1)U_{11}/W_1 \times$$

$$\times \left\{ \sin k_1(L - |y|) - \frac{W_1(1 - b_1)(1 + c_2)}{W_2(1 - b_2)(1 + c_1)} \sin k_2(L - |y|) - d_1\left[\cos k_1(L - |y|) - \cos k_2(L - |y|)\right] \right\}.$$

$$(8.46)$$

The field in the far region taking a mirror image into account is

$$E_\theta = \frac{60k(1 + c_1)U_{11}\exp(-jkr)\sin\theta}{W_1 r}\left[\frac{\Theta_1(\cos\theta)}{k^2\cos^2\theta - k_1^2} - \frac{\Theta_2(\cos\theta)}{k^2\cos^2\theta - k_2^2}\right], (8.47)$$

where

$$\Theta_i(\cos\theta) = k_i e_i\left[\cos(kL\cos\theta) - \cos k_i L\right] + d_1[k\cos\theta\sin(kL\cos\theta) - k_i\cos k_i L],$$

$$e_1 = 1, e_2 = \frac{W_1(1 - b_1)(1 + c_2)}{W_2(1 - b_2)(1 + c_1)}.$$

An effective length of asymmetric radiator is

$$h_e = \frac{k_2(1 - \cos k_1 L) - e_2 k_1(1 - \cos k_2 L) - d_1(k_2\sin k_1 L - k_1\sin k_2 L)}{k_1 k_2\left[\sin k_1 L - e_2\sin k_2 L - d_1(\cos k_1 l - \cos k_2 L)\right]}. (8.48)$$

Thus, the calculation of the folded radiator with nonzero boundary conditions is divided into two stages. First, the current distribution along the antenna wires is determined using the theory of asymmetric lines, afterwards electrical characteristics of the antenna are calculated. In order to calculate the far field, the total current of an antenna is used. Input impedance is calculated by the method of induced electromotive force, or by solving integral equation. Coefficients W_i, b_i, c_i, k_i depend on the inductive reactances $X_{ik} = \omega p_{ik}/c^2$ per unit length, where p_{ik} are the potential coefficients, which are determined by a method of an average potential (for example, by method of Howe), in accordance with the actual location of the antenna wires.

Practically important special cases, when the surface impedance on one of the wires of the folded radiator is equal to zero, are of particular interest. Main characteristics of folded antennas with a gap, if one or another wire is purely metallic, are given in Table 8.2. One must note that in the calculation of the difference $k_1^2 - k^2$ it is necessary to expand it into the series of Maclaurin. If the radiator is made up of two identical wires ($Q_1 = Q_2$, $a_2 = a_4$), then

Table 8.2: Characteristics of folded radiators with a gap.

Characteristics	$Q_2 = 0$	$Q_1 = 0$
k_1	k	k
k_2	$k\sqrt{1 + \dfrac{Q_1 X_{22}}{X_{11}X_{22} - X_{12}^2}}$	$k\sqrt{1 + \dfrac{Q_2 X_{11}}{X_{11}X_{22} - X_{12}^2}}$
$k_1^2 - k^2$	$Q_2 \dfrac{k^2}{X_{22}}$	$Q_1 \dfrac{k^2}{X_{11}} - Q_1^2 \dfrac{k^2 X_{12}^2}{Q_2 X_{11}^3}$
X_{il}	$-\dfrac{k_2 Q_1}{k_2^2 - k^2} \times \dfrac{F_1 + F_2}{F_3 + F_4} \cot kL,$ where $F_1 = 1 + \dfrac{(X_{22} - X_{12})\cos k_2 L}{X_{12}\cos k_2 l},$ $F_2 = d_1 \tan kL \left\{ 1 - \dfrac{k_2 k Q_1 \sin k_2 L}{\left[(k_2^2 - k^2)(X_{12} - X_{11}) + k^2 Q_1\right]\sin kL} \right\}$ $F_3 = \dfrac{(X_{22} - X_{12})\sin k_2 L}{X_{12}\sin kL},$ $F_4 = d_1 \tan kL \left\{ \dfrac{k_2 k Q_1 \sin k_2 L}{\left[(k_2^2 - k^2)(X_{12} - X_{11}) + k^2 Q_1\right]\sin kL} \right\}$	$-X_{11}\dfrac{1 + d_1 \tan kL}{k(1 - d_1 \tan kL)}\cot kL$
d_1	$\dfrac{1 + \dfrac{X_{22} - X_{12}}{k_2 k Q_1 X_{12}}\left[(k^2 - k_2^2)X_{11} + k^2 Q_1\right]\dfrac{\sin k_2 L}{\sin kL}}{1 + \dfrac{\left[(k_2^2 - k^2)X_{11} - k^2 Q_1\right]\cos k_2 L}{\left[(k_2^2 - k^2)(X_{12} - X_{11}) + k^2 Q_1\right]\cos kL}}$	$-\dfrac{(k_2^2 - k^2)(X_{11} - X_{12})^2}{k_2 k Q_2 X_{11}}\cot kL$
e_2	$\dfrac{X_{12} - X_{22}}{k_2 k Q_1 X_{12}}\left[(k_2^2 - k^2)(X_{12} - X_{11}) + k^2 Q_1\right]$	$-\dfrac{(k_2^2 - k^2)(X_{11} - X_{12})^2}{k_2 k Q_2 X_{11}}$

$$k_1 = \sqrt{k^2 + \omega Q_1(\beta_{11} + \beta_{12})} = k\sqrt{1 + Q_1/(X_{11} + X_{12})},$$

$$k_2 = \sqrt{k^2 + \omega Q_1(\beta_{11} - \beta_{12})} = k\sqrt{1 + Q_1/(X_{11} - X_{12})},$$

whence $b_1 = c_1 = 1$, $b_2 = c_2 = -1$, $d_1 = 0$, i.e., expressions (8.40) take the form

$$u_1 = U_{11}\cos k_1 z + U_{12}\cos k_2 z + j\left(W_1 I_{11}\sin k_1 z + W_2 I_{12}\sin k_2 z\right),$$

$$i_1 = I_{11}\cos k_1 z + I_{12}\cos k_2 z + j\left(\frac{U_{11}}{W_1}\sin k_1 z + \frac{U_{12}}{W_2}\sin k_2 z\right),$$

$$u_2 = U_{11}\cos k_1 z - U_{12}\cos k_2 z + j\left(W_1 I_{11}\sin k_1 z - W_2 I_{12}\sin k_2 z\right),$$

$$i_2 = I_{11}\cos k_1 z - I_{12}\cos k_2 z + j\left(\frac{U_{11}}{W_1}\sin k_1 z - \frac{U_{12}}{W_2}\sin k_2 z\right). \quad (8.49)$$

This means that in this particular case, irrespective of the boundary conditions for the currents and voltages their components with the propagation constant k_1 are equal in magnitude and opposite in sign (anti-phase wave).

Accordingly, the input impedance of the folded radiator with a gap (see Fig. 8.12a) can be presented as an aggregate of input impedances of two-wire line and monopole:

$$X_{il} = -W_m \cot k_1 L + \frac{W_L}{4} \tan k_2 L, \tag{8.50}$$

where $W_m = W_1/2 = \dfrac{k_1}{2kc(\beta_{11} + \beta_{12})} = \dfrac{k_1}{2kcC_{11}}$ is the wave impedance of the impedance linear radiator consisting of two parallel wires, and

$W_l = 2W_2 = \dfrac{2k_2}{kc(\beta_{11} - \beta_{12})} = \dfrac{k_2}{kc(C_{12} + C_{11}/2)}$ is the wave impedance of

an impedance long line also consisting of two wires located symmetrically relatively surface of zero potential (ground). Magnitudes C_{11} and C_{12} in these expressions are partial capacitances.

For the folded radiator with the second wire shorting to ground (see Fig. 8.12b) instead of the third boundary condition (8.41) we have

$$u_2(L) = 0. \tag{8.51}$$

Therefore, instead of the third equation of the system (8.42), we obtain

$$I_{11} = -jd_2 \frac{U_{11}}{W_{11}}, \tag{8.52}$$

where

$$d_2 = -\cot k_1 L \left[1 + \frac{b_2(1-b_1)\cos k_2 L}{b_1(1-b_2)\cos k_1 L} \right] \left[1 - \frac{W_2 b_2(1+c_1)\sin k_2 L}{W_1 b_1(1+c_2)\sin k_1 L} \right]^{-1},$$

and d_2 will take the place of the coefficient d_1 in expressions for electrical characteristics of the radiator. When $Q_2 = 0$, coefficient d_2 is equal to $d_2 = -\cot kL$. When $Q_1 = 0$,

$$d_2 = -\cot kL \left[1 + \frac{(X_{11} - X_{22})\cos k_2 L}{X_{12}\cos kL} \right] \left[1 - \frac{k_2 kQ_2 X_{11}\sin k_2 L}{(k_2^2 - k^2)X_{12}(X_{11} - X_{12})\sin k_1 L} \right]^{-1}.$$

When $Q_1 = Q_2, a_2 = a_4,$

$$Y_{il} = \frac{1}{W_l \tan k_2 L} - \frac{1}{4W_m \cot k_1 L}. \tag{8.53}$$

When $Q_1 = Q_2 = 0$, i.e., $k_1 = k_2 = k$, equalities (8.50) and (8.53) are transformed into expressions (8.4) and (8.5).

As an example, in Fig. 8.13 the model of the impedance folded radiator is presented. One wire of this radiator is made in the form of a rod with a ferrite coating (relative magnetic permeability of the coating is 10), and the other wire is made in the form of a metal tube. Dimensions of a model are given in millimeters.

The calculated curves and experimental values of active R_A and reactive X_A components of the input impedance of the impedance folded radiators are given in Figs. 8.14 and 8.15. In different variants the generator is connected to different wires, the second wire is connected or not connected to the ground, wires of different diameter are used. The coincidence of the calculated and experimental results is quite satisfactory. As is seen from the figures, the radiator characteristics are substantially changed, if one or other wire is excited. Using slowing coating allows to decrease resonant frequencies by a factor of 2–2.5.

Comparing the characteristics of impedance and metal folded antennas, it is easy to see that both impedance antennas and metal ones allow to significantly reduce the resonant frequencies in comparison with linear antennas of the same height. In the case of impedance antennas, there is an additional degree of freedom for choosing a resonant frequency, which depends on the dielectric permittivity and magnetic permeability of the coating material.

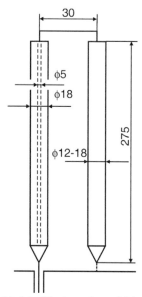

Fig. 8.13: Model of the impedance folded radiator.

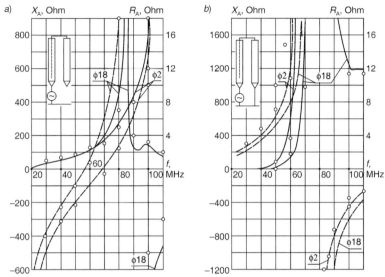

Fig. 8.14: Input impedance of the impedance folded radiator with excited impedance wire: a—with a gap, b—with a shorting to the ground.

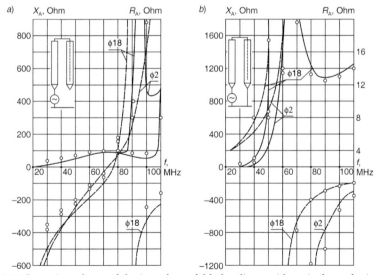

Fig. 8.15: Input impedance of the impedance folded radiator with excited metal wire: a— with a gap, b—with a shorting to the ground.

9

Multi-folded Antennas, Perpendicular to Metal Surface

9.1 Principle of operation and method of calculation

As is shown in Section 8.1, asymmetric folded radiator consisting of two parallel wires, upper ends of which are connected with each other, combines the functions of radiation and matching. In folded radiator with a gap, length of which is less than a quarter of wave length, capacitive impedance of a linear radiator (monopole) is compensated by inductive impedance of short-circuited long line. In the folded radiator with shorting to a ground, the long line, connected in parallel with input impedance of linear antenna, transforms its resistance.

Multi-folded radiator (Fig. 9.1) consist of several folded radiators connecting in series with each other and gives more opportunities in order to obtain new circuits and connections (series and parallel ones). This radiator is a group of parallel wires connected in pairs on top and bottom so that to form a system of coupled and connected in series elongated loops (of long lines). In the particular case when the number of wires is two, this antenna becomes a folded radiator.

If the transverse dimensions of multi-folded radiator are small in comparison with its height L and the wave length λ, then, as is shown in the article [43], devoted to a research of electromagnetic oscillations in systems of parallel thin wires, the current in each wire of such a system can be divided into in-phase and anti-phase components, and the entire system may be reduced to an aggregate of linear radiator and non-radiating long lines.

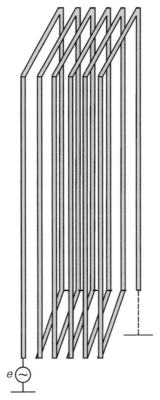

Fig. 9.1: Multi-folded radiator.

The calculating method of such antennas may be considered using an example of a two-folded radiator with a gap (Fig. 9.2).

At first, we must divide the two-folded antenna onto a radiator and long lines. For that, into the gap between the free end of the antenna and the ground we include two generators of the current that are equal in magnitude (mJ) and opposite in sign (Fig. 9.3). Here J is the current of the main generator. The main generator is divided onto two generators of the current that are identical in direction and different in magnitude, with currents mJ and $(1-m)J$. The total current of the generator as a result of such operation is not change; the total current in the gap is zero as before.

According to the superposition principle a voltage at point A is equal to a sum of voltages, produced by all generators. Therefore, as it is shown in Fig. 9.3, one can divide the circuit of the two-folded antenna onto two circuits, with two generators in each one. The voltages at point A, created in each of these circuits, are calculated and summed. In the first auxiliary circuit the generators are identical and connected in series. Therefore, the voltage between the point A and the ground is

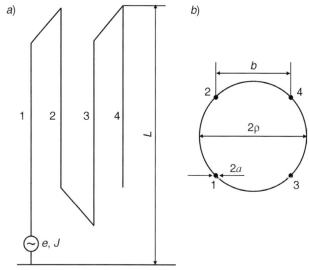

Fig. 9.2: Two-folded radiator with a gap: a—circuit, b—cross section.

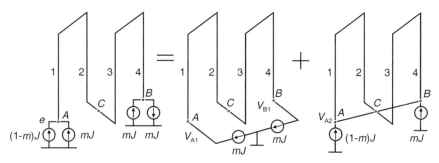

Fig. 9.3: To a calculation of the two-folded radiator with a gap.

$$V_{A1} = V/2 = 0.5 \, mJZ_{l'}$$ (9.1)

where $V = V_{A1} - V_{B1}$ is the input voltage, and Z_l is the impedance of an complicated long line. More precisely, there are two coupled lines with the same distance b between wires and the equal wave impedances $W_1 = 120 \ln (b/a)$. One of these lines is a load for the other line. In the first approximation one can assume that it is united two-wire line, bended at an angle 180° in the middle. If to use the theory of electrically coupled lines, W_1 in this expression will be replaced by the value $W_2 = \dfrac{W_3 W_4}{W_4 - W_3}$, where $W_3 \sim 80 \ln b/a$, $W_4 = 240 \ln [b/(a\sqrt{2})]$. It is easy to make sure that for small radius of wires ($b \gg a$)

$$W_2 \approx W_1 \approx 120 \ln (b/a).$$

The points A and B in the second auxiliary circuit are connected with each other as equipotential points. It means that $m = 1/2$, if the wires diameters are identical. The second circuit is the folded radiator, in which each "conductor" consists of two parallel wires (1 and 4, 2 and 3 respectively). This circuit can also be divided into two ones: the four-wire line of length L, shorted at the end, and the asymmetric radiator (monopole) of height L. The wave impedance W_l of the line is equal to $W_5 = 60 \ln [b/(a\sqrt{2})]$. Equivalent radius of the monopole with four wires (of radius $b/\sqrt{2}$), located along the cylinder generatrices, is equal to $a_e = \sqrt[4]{ab^3 \sqrt{2}}$. Input impedance of the second circuit is

$$Z_{A2} = Z_m (a_e) + jm^2 W_l \tan kL.$$

This impedance is calculated by means of the same procedure of dividing the initial circuit onto two circuits (monopole and long line) and summing the voltages. For the voltage between point A and the ground in the second circuit one can write

$$V_{A2} = J[Z_m(a_e) + j0.25 \, W_5 \tan kL], \tag{9.2}$$

i.e., the input impedance of the entire antenna

$$Z_A = Z_m (a_e) + j0.25 \, W_2 \tan 2kL + j0.25 \, W_5 \tan kL. \tag{9.3}$$

This result is illustrated by Fig. 9.4, which shows the impedances of the antenna and its components. The input impedance of the considered antenna is a series connection of the monopole and two lines with length L and $2L$ shorted at the end. The wave impedances of these lines are close to values

$$W(n) \approx (120/n)\ln (b/a), \tag{9.4}$$

where n is the number of wires in each "conductor" of the line.

From (9.4) and Fig. 9.4 it follows that the radiation resistance of the two-folded antenna with a gap is equal to the radiation resistance of the monopole of the same height. The reactive component of the input impedance has additional resonances, and the first parallel resonance is caused by a parallel resonance of the long line with length $2L$, i.e., its frequency is half the frequency of the first series resonance of an ordinary monopole with the same height. The frequency of the first series resonance of the antenna is even smaller (but not necessarily two times).

Due to increase of the wires quantity and a corresponding increase of the length of a total antenna wire, the number of resonances in a concrete

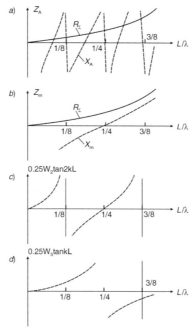

Fig. 9.4: Impedance of two-folded antenna (a) and of its components: monopole (b), long line of length 2L (c), long line of length L (d).

frequency range increases. For example, if the number of folded radiators connecting in series with each other is equal to $N = 2^n$, then for an antenna with a gap we obtain similarly to (9.3)

$$Z_A = Z_m (a_e) + j0.25 \sum_{m=0}^{n} W (2^m) \tan (NkL/2^m), \qquad (9.5)$$

where $a_e = \sqrt[2N]{2Na\rho^{2N-1}}$ is the equivalent radius of the monopole, consisting of 2N wires, which are located along the generatrices of the cylinder with the radius ρ (if N grows, the equivalent radius tends to ρ). The frequency of the first parallel resonance of the antenna (i.e., of its second resonance) is N times lower than the frequency of the first series resonance of an ordinary monopole with the same height. Such a character of the input impedance allows, firstly, to use multi-folded antenna in the range of longer waves, and secondly, when it is necessary, to tune the antenna onto several frequencies.

If N-folded antenna with shorting to ground has the wires of identical diameters (see Fig. 9.1, dotted line), its input admittance is

$$Y_A = \frac{1}{j120 \ln(b/a) \tan NkL} + \frac{1}{4Z_{N/2}}, \qquad (9.6)$$

where $Z_{N/2}$ is the input impedance of the $N/2$-folded antenna with a gap and with "conductors" from two wires. This result generalizes expression (8.5) and gives a similar result.

In the case of odd number of antenna wires the calculation becomes more complicated. For example, a radiator of three wires (Fig. 9.5) may be divided onto a three-wire line and a monopole of a height L (Fig. 9.6). Potentials of all wires in each cross section of a second circuit (monopole) must be the same. So, in accordance with (8.8) the magnitude m depends on the capacitance relations of two antenna branches. The right branch consists of two wires, and its capacitance is twice as much. Therefore, from here it is follows that $m = 2/3$. One can show, using the theory of electrically

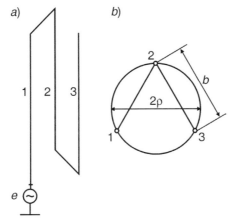

Fig. 9.5: Three-wire antenna: *a*—circuit, *b*—cross-section.

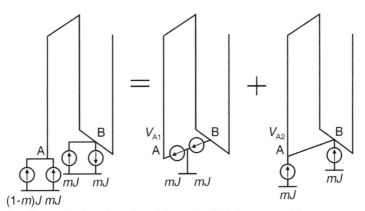

Fig. 9.6: To the calculation of three-wire folded antenna with a gap.

coupled lines that an impedance of a three-wire line of the identical three wires (of a first circuit) is equal to

$$Z_I = j80 \ln(b/a)\frac{1}{\cot 2kL + \cot kL}. \tag{9.7}$$

In this circuit there are only anti-phase currents, and their sum is equal to zero in each cross section. Sum of potentials in an arbitrary cross section also is zero, i.e.,

$$V_{A1} = -2V_{B1} = 2V/3,$$

where $V = V_{A1} - V_{B1}$ is the voltage at the input of the long line. From here the impedance of the three-wire antenna is

$$Z_A = \frac{V_{A1} + V_{A2}}{J} = Z_m(a_e) + j80 \ln(b/a)\frac{1}{\cot 2kL + \cot kL}. \tag{9.8}$$

Equivalent radii of the three-wire monopole and the three-folded radiator are equal accordingly to $a_e = \sqrt[3]{ab^2}$ and $a_e = \sqrt[6]{6ab^5}$. If the number of loops is $N = 3 \cdot 2^n$, then

$$Z_A = Z_m(a_e) + j0.25 \sum_{m=0}^{n} W(2^m) \tan(NkL/2^m) + j0.33 W(2^n) \frac{1}{\cot 2kL + \cot kL}. \tag{9.9}$$

The value of a_e is given earlier.

9.2 Electrical characteristics of multi-folded radiators

The first section of this chapter is devoted to the method of analysis of multi-folded antennas and to their input impedances. Here we shall briefly talk about other properties of these radiators.

The directional pattern of multi-folded antenna does not differ from the directional pattern of an ordinary monopole, since fields of long lines in a far zone may be neglected if distances between wires are small. In calculating a loss resistance R_{gA} of multi-folded antennas in the ground there is a need to determine loss resistance R_{gl} for lines of complex shape. Figure 9.7a shows for example a two-wire line of length 2L, which in the middle is bent at an angle of 180°. Losses of such line in a ground do not differ from losses of a line shown in Fig. 8.11b because currents in the ground between projections of the wires 2 and 3 are practically absent. The current between

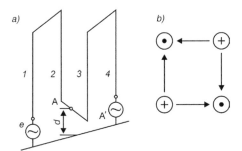

Fig. 9.7: Losses in the ground for lines of complex shape (a) and with two wires in "conductor" (b).

the wires 2 and 3 flows mainly along a connecting bridge AA', especially if the distance between these wires is less than a distance d between the connecting bridge and the ground.

The loss resistance of a long line, where each "conductor" consists of n wires, is smaller by a factor n^2 than the loss resistance of ordinary line. This situation is illustrated by Fig. 9.7b for the case when $n = 2$. The current of each wire of a "conductor" flows into the n directions to n wires of the other "conductor", thereby forming a system of n^2 resistances, included in parallel. Each resistance is equal to the resistance R_{gl0} of a sector between single wires. Thus, the loss resistance in the ground of multi-folded radiator with a gap is equal to

$$R_{gA} = R_{ge} + \frac{1}{4} \sum_{m=0}^{n} \frac{R_{gl0}}{n_m^2}, \qquad (9.10)$$

where n_m is number of wires in each "conductor" of line m. Values R_{gA} for some antennas are given in Table 9.1. The table demonstrates that if the transverse dimension of an antenna is the same, increasing number of folded radiators affects weakly the losses in the ground.

Further we consider losses in the antenna wires caused by a skin effect. As is known, the surface resistance of a round copper wire per one meter is

$$R_{\sim} = \sqrt{f}/24a, \qquad (9.11)$$

where f is a frequency, in megahertzs, a is a wire radius, in millimeters. The resistance of a steel wire is 2.3 times as much. The surface resistance causes longitudinal attenuation of electromagnetic waves. Therefore, the propagation constant of a wave along the wire as well as wave impedances of a long line and a monopole become complex quantities. If to include these impedances into expressions for input impedances of antennas, one can find the input impedances of antennas with allowance of the losses from skin effect.

Table 9.1: Loss resistance of multi-folded radiators with a gap.

Type of radiator	Loss resistance in a ground R_{gA}	Loss resistance in wires R_{wA}
folded	$R_{ge} + 0.25\, R_{gl0}$	$\dfrac{R\_L}{\sin^2 2kL}\left(1 - \dfrac{\sin 4kL}{4kL}\right)$
two-folded	$R_{ge} + 0.313\, R_{gl0}$	$\dfrac{2R\_L}{\sin^2 4kL}\left(1 - \dfrac{\sin 8kL}{8kL}\right)$
four-folded	$R_{ge} + 0.328\, R_{gl0}$	$\dfrac{4R\_L}{\sin^2 8kl}\left(1 - \dfrac{\sin 16kL}{16kL}\right)$
linear	R_{ge}	$\dfrac{R\_L}{2\sin^2 2kL}\left(1 - \dfrac{\sin 2kL}{2kL}\right)$

The imaginary additive to the propagation constant in the first place increases the magnitude of the active component R_A, since the latter is small in comparison with the reactive component X_A everywhere except the vicinity of resonances. An addition to a given above radiation resistance R_Σ is a sought value of loss resistance RWA in the wires.

In Table 9.1 values R_{WA} are given for several variants of multi-folded radiators with a gap, when attenuation in the wires is weak. It is believed that losses are small when the wires radii are small compared with the distances between them, i.e., one must neglect the proximity effect and the corresponding redistribution of the current over the wire cross section. For comparison, the table shows the loss resistance in the wires of a monopole.

The table shows that the losses in the wires cause the appearance of additional maxima on the curve plotted for the active component of an input impedance when $kL = (2m + 1)\pi/2$ (m is a natural number), i.e., near the parallel resonance of long lines. In these frequency bands it is impossible to ignore the losses to a skin effect. At low frequencies ($kL \ll 1$) loss resistance increases proportionally to the number and length of wires.

An analysis permits to find the input admittance of the N-folded radiator with the ground connection of the unexcited wire. With allowance for the losses in the ground and in the wires this admittance is equal to

$$Y_A = \frac{1}{j120\ \ln(b/a)\tan NkL + R_{gl} + R_{wl}} + \frac{1}{4Z_{N/2}}, \qquad (9.12)$$

where $Z_{N/2}$ is the input impedance of the N/2-folded antenna with a gap, consisting of two-wire "conductors" in view of losses; R_{gl} is the loss resistance in the ground; $R_{wl} = \dfrac{NR\_L}{\cos^2 NkL}$ is the loss resistance in the wires of a two-wire long line.

Calculation and experimental verification confirmed the rightness of the obtained results. Figures 9.8 and 9.9 show the input impedances of two-folded and four-folded radiators with a gap in different frequency bands. The experimental values of the resonant frequencies in the figures are shifted in comparison with the calculated results in the direction of lower frequencies. This small shift is caused by the fact that the calculation did not take into account the length of the horizontal connecting bridges between the wires.

Further we pass to Q-factor. Q-factor (quality) is an important electrical characteristic of the antenna. It defines in particular the frequency band, within which one may obtain without change of the tuning a given level of matching an antenna with a cable. Q-factor characterizes the rate of changing the antenna input impedance as a result of the influence of various external factors and can be used to quantify the sustainability of the antenna tuning, if sustainability is understood as a preservation of a tuning (stability). From this standpoint, the higher the quality factor, the worse the stability, and vice versa.

The parameter Q is a value similar to the quality factor of the resonant circuit. It is calculated at the point of the series resonance of antenna in accordance with the formula

$$Q_i = \frac{\omega_i}{2R_{Ai}} \frac{dX_A}{d\omega}\bigg|_{\omega=\omega_i} = \frac{(kL)_i}{2R_{Ai}} \frac{dX_A}{d(kL)}\bigg|_{kL=(kL)_i}, \tag{9.13}$$

where R_{Ai} is the active component of the input impedance on the frequency f_i, in the vicinity of which R_{Ai} assumed constant.

The expressions for calculating the electrical length $(kL)_i$ and Q-factor of four-folded radiator with a gap are given as an example in Table 9.2 in the first four points of the series resonance. The table also indicates for the comparison the electrical length and the quality factor of two other radiators: a quarter-wavelength radiator and a short linear radiator with a matching device. The table uses the following designations: W_m is the wave impedance of two-wire long line; $\Delta = \sqrt{W_l/(4W_m + 5W_l)}$, $p = \lambda/(4L)$.

9.3 Using multi-folded radiators in compensation devices

Multi-folded antennas can play an important role in solving a problem of creating weak fields near a transmitting antenna. This problem is relevant when near the transmitting antenna is a body that is sensitive to an electromagnetic field, and this body must be protected from this field, without screening it from an external space. This situation arises when solving a problem of electromagnetic compatibility and when protecting people from irradiation. The first problem is that exposure on nearby

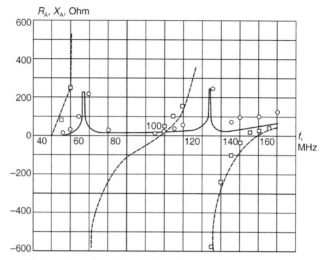

Fig. 9.8: Input impedance of two-folded radiator with a gap.

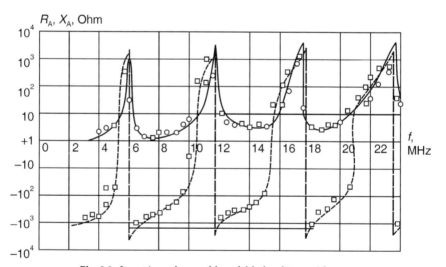

Fig. 9.9: Input impedance of four-folded radiator with a gap.

appliances can disrupt their normal operation, lead to spontaneous switch-on and switch-off of them, regime change, etc. Frequency diversity does not always satisfactorily solve this problem.

The second problem is connected with the fact that a personal cell phone is located next to the user's head during a conversation, and its transmitter irradiates sensitive human organs (brain, eyes). In today's phones, this absorbs about half of a total radiated energy [44, 45], turning

Table 9.2: Expressions for calculating electrical length and quality of four-folded and linear radiators.

Radiator	Resonance number	Electrical length	Q-factor
four-folded	1	$\dfrac{1}{4}\tan^{-1}\left(9.63\dfrac{W_m}{W_l}-0.604\right)$	$\dfrac{1}{R_{A1}}\left(-0.94W_m+0.38W_l+18.2\dfrac{W_m^2}{W_l}\right)$
"	2	$\dfrac{1}{2}\tan^{-1}\left(8\dfrac{W_m}{W_l}-0.5\right)$	$\dfrac{1}{R_{A2}}\left(0.56W_l+6.3\dfrac{W_m^2}{W_l}\right)$
"	3	$\dfrac{1}{4}\tan^{-1}\left(1.66\dfrac{W_m}{W_l}-0.1\right)$	$\dfrac{1}{R_{A3}}\left(0.5W_m+1.14W_l+1.62\dfrac{W_m^2}{W_l}\right)$
"	4	$(\pi-\delta)/2$	$\dfrac{1}{R_{A4}}\left[0.79\left(1+\Delta^2/4\right)W_m+\right.$ $\left.+\left(1.03+3.34\Delta^2+0.196/\Delta^2\right)W_l\,\right]$
quarter-wavelength	1	$\pi/2$	$W_m/93$
short with a matching device	1	$\pi/(2p)$	$\pi W_m/[4pR_A\sin^2(\pi/2p)]$

it into heat. It is necessary to reduce a quantity of electromagnetic energy that the head absorbs.

An idea of protection from electromagnetic field by screening, i.e., by applying a shadowing effect, appears obvious, but it is practically not realizable. The near field has no ray structure. Therefore, the shadow formed behind the metal screen extends only over a distance close to the screen dimensions. For example, in order to protect the user's head from radiation, it is necessary to apply a screen substantially larger than the transverse dimension of the cellular phone. For similar reasons, one must to avoid an absorber, i.e., dielectric shield. The obvious drawback of using large screens and absorbers is the distortion of an antenna pattern.

More acceptable results are given by another principle—by the principle of mutual compensation of fields created by different radiating elements in a given area. This method is used to protect equipment near power lines. In a special variant of this method for creating a compensating field and a "dark spot" an additional radiator is used, which is excited by the same source of energy that creates a field of the main radiator [46]. The method is based on a rapid decrease of the near field with increasing the distance from the radiating element. The field of a small radiator can in some region have the same value as the field of a large radiator located further from this region. In this case, a small field, as a rule, practically does not affect the shape of the directional pattern, since near the antenna the radiated field does not consist of rectilinear rays but gently rounds the "dark spot".

The structure of the offered antenna [47] is shown in Fig. 9.10. It is an asymmetrical dipole, which is excited by a generator 1 at a feed point 2. The lower arm of the dipole is a metal plate 3, to bottom edge of which a small plate 4 is attached at an acute angle. The upper arm is a multi-folded structure 5, which is excited at a mid-point 6. It is fabricated in the shape of three-dimensional structure that protrudes to the direction of the user's head. A dielectric plate 7 is inserted inside this structure. This structure creates anti-phase currents in the antenna and hence small auxiliary fields, which compensates the main field at a certain point (compensation point) and forms around this point an area of a weak field (a dark spot).

This arm has a small height. Its contribution to the antenna's radiation is small, but it allows to match the antenna to a cable or a generator. The reactive component of its impedance compensates the reactive component of the lower arm, providing series resonances at operating frequencies. Its complex structure has many degrees of freedom, including number of sections, their dimensions, the width and thickness of the wires, types and magnitudes of concentrated loads. That permits to change the input reactance of the antenna in wide limits and to provide operation at few bands of frequencies.

The input impedance of an asymmetrical dipole, consisting of two different arms, in the first approximation, is equal to half the sum of input impedances of two symmetrical radiators: one with arms identical to the lower arm of the asymmetrical dipole, and other radiator with arms identical to its upper arm [48]. This relation is exact for the input reactance of the antenna and has an approximate nature for its resistance. From the above it follows that

$$X_A = X_{A1} + X_{A2}, \quad R_A \approx R_{A1} + R_{A2}, \tag{9.14}$$

where X_{A1} and R_{A1} are the reactive and resistive components of input impedance of the radiator, shown in Fig. 9.11a, X_{A2} and R_{A2} are corresponding components of the radiator, shown in Fig. 9.11b.

The lower arm of the offered antenna is a monopole, realized as a wide metal plate with length L_1 and width d. The input impedance of the monopole in the first approximation is equal to

$$Z_{A1} \cong R_{A1} - jW_1 \cot kL_1. \tag{9.15}$$

Here $W_1 = 30\Omega$ is the wave impedance of the monopole, where $\Omega = 2\ln(2L_1/a_{e1}) \approx 1/\chi$ is the parameter of the theory of linear antennas, R_{A1} at frequencies near the first series resonance is close to 40 Ohm. If $d \approx L_1/2$, then $a_{e1} \approx L_1/(2\pi)$, and $\Omega = 2\ln(4\pi) \approx 5$.

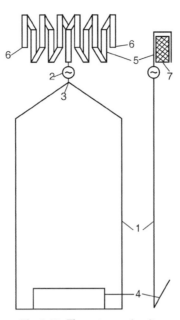

Fig. 9.10: The antenna circuit.

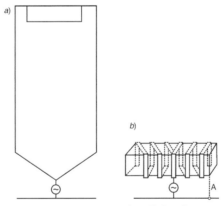

Fig. 9.11: Components of a cellular phone.

The upper arm of the antenna in the first approximation may be considered as a multi-folded radiator, mounted on a metal plate (see Section 9.2). The current in each wire can be divided into in-phase and anti-phase components. This system is reduced to an aggregate of a radiating linear antenna (monopole) and few non-radiating long lines. The multi-folded radiator can have built either with shorting to ground or with a gap (see Fig. 9.11b). In the latter case the input impedance of the multi-folded radiator is a series connection of the monopole and the two-wire long lines, shorted at their ends. The expressions for the input impedances $Z_{A2}^{(n)}$ of these antennas with a different number n of wires are presented in Section 9.2.

As is shown in Section 9.2, the radiation resistance of this antenna is equal to the radiation resistance of a monopole with the same height. The reactive component of its input impedance has additional resonances, and the first parallel resonance is defined by a line of length $SL/2$, i.e., its frequency is smaller approximately by a factor of $S/2$ in comparison with the frequency of the first series resonance of an ordinary monopole with the same height. The frequency of the first series resonance of the multi-folded antenna is still smaller, approximately twice. Such type of the input impedance simplifies selecting dimensions of a structure, whose series resonances coincide with the given operating frequencies. Thus, in this case the monopole's effective length is

$$h_e = (1/k)\tan(kL_1/2) + L_2/2, \tag{9.16}$$

where L_2 is the length of the upper arm. If, for example, $kL_1 = \pi/2$, and $L_2 = L_1/4$, then $h_e = 1/k + L_1/8$. Accordingly, the radiation resistance of the antenna is

$$R_\Sigma = 20k^2 h_e^2 = 20(1 + \pi/16)^2 \approx 29. \tag{9.17}$$

From that it is clear that the radiation resistance of the monopole (of the metal plate) determines mainly the antenna's resistance on the whole, i.e., the latter is close to 30 Ohm, if the operating frequency is close to the frequency of the first series resonance. Contribution of the upper arm to the radiation resistance is small as compared with the contribution of the lower arm. But the reactance of the monopole is close to zero, if the plate length is equal to a resonant length, i.e., if it is close to $\lambda/4$. Therefore, to ensure matching with a cable or generator, the upper arm should have a series resonance at the first operating frequency. Accordingly, the reactance of the upper arm of a multi-band antenna should compensate the monopole reactance at other operating frequencies.

The directional pattern of antenna in the horizontal and vertical planes do not differ from a typical directional pattern of monopole, since the radiation of the long lines can be neglected, provided the distances between the wires are small. The analysis of the considered antenna demonstrates that its structure firstly allows to obtain the series resonances at required frequencies. Secondly, the considered antenna contains a radiating element of large length, close to $\lambda/4$, and that permits providing an effective radiation and reception of signals.

If the multi-folded antenna is shorted to ground at points A, its input admittance is equal to

$$Y_{A2}^{(S)} = 1/[4Z_{A2}^{(S/2)}] + 1/[j120 \ln(b/a) \tan (SkL/2)], \tag{9.18}$$

where $Z_{A2}^{(S/2)}$ is input impedance of an antenna with a gap, whose conductors consist of two wires. Therefore, in this case the input impedance is a parallel connection of a $S/2$-folded radiator with a gap and a close-end line of length $SL/2$. Such type of the input impedance complicates selecting dimensions of a structure.

In order to produce a small auxiliary field with the aim to compensate the main field in the user's head without changing the far field, it is expedient to use the multi-folded structure, which protrudes toward the user's head. This structure allows to create an anti-phase field in the near region of the antenna and to nullify the total field at a compensation point, located not far from the neighboring edge of the user's head. Around the mentioned point an area of a weak field (a dark spot) is created. We shall demonstrate this effect by means of the folded radiator with a gap in a point A, which is located along a plane passing through the compensation point and placed on the side of the phone housing near to the head.

The folded radiator consists of two parallel wires. If to connect two current generators of equal magnitude $J/2$ and opposite directions to the bottom of the right wire in parallel with each other and to divide also the main generator into two parallel generators, equal in magnitude (to $J/2$) and coinciding in direction, then as a result, voltages and currents of this circuit do not change. According to the superposition principle, the voltage and the currents at each point are equal to the sum of the voltages and the currents, produced by all generators. Therefore, as shown in Fig. 9.12, one can divide the considered circuit onto two circuits with two generators in each and then calculate and sum the currents in any wire, created in each circuit. The left-hand circuit is a linear radiator (monopole), the right-hand circuit is a two-wire long line.

The current in each wire consists of an anti-phase current of the line and the component of the in-phase current in the linear radiator. In Fig. 9.13 the currents' distribution along the wires is presented: the distribution of the in-phase currents $J_1^{(in)} = pJ$ and $J_2^{(in)} = mJ$ along wires 1 and 2 of the monopole (*a*), the distribution of the anti-phase current $J_1^{(an)} = mJ$ and $J_2^{(an)} = mJ$ along the wires of close-end line (*b*) and the total currents' distribution (*c*).

In-phase currents in both wires are distributed by sinusoidal law; the ratio of their currents depends on the wire capacitances, i.e., for identical wires the currents are equal. The anti-phase currents are the same in magnitude, but opposite in sign. They are distributed by cosine law. In the first wire the currents are added, since they have the same sign. In the second wire they are opposite in sign, and under these conditions the dipole moment of negative current is greater due to cosine distribution. This gives a possibility for compensation of the fields, created by the wires. The total current of the second wire is less than the total current of the first wire. But since the second wire is located nearer to the compensation point, then,

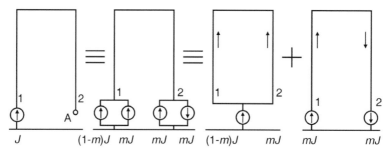

Fig. 9.12: Division of the folded radiator into two circuits.

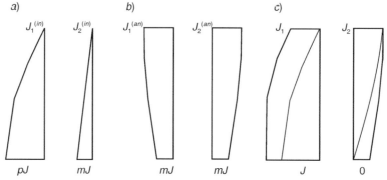

Fig. 9.13: The distribution of in-phase (a), anti-phase (b) and total (c) currents.

although the wires' currents are different, their fields at the compensation point are the same in magnitude.

This effect is more pronounced if, firstly, the folded radiator is replaced by a multi-folded structure, and, secondly, if a dielectric plate is placed between the wires. In this case the field of the first wire is attenuated quicker with every spacing. As a result, firstly, one can decrease this distance, i.e., decrease the thickness of phone housing. Secondly, the electric lengths of transmission lines increase, i.e., one can decrease geometric lengths of lines, for example, to excite the multi-folded structure in the middle of its width and not at a side point.

The additional anti-phase current is produced by the small plate 4 (see Fig. 9.10).

It should be emphasized that the structure without a connecting bridge between the upper ends of the wires is divided into a monopole and an open-end two-wire line. In this case, the in-phase and anti-phase currents of the second wire are equal in magnitude, i.e., the total current of the second wire is zero, and hence the possibility of compensating the field of first wire is absent.

It is useful to compare the characteristics of the proposed antenna with the characteristics of a symmetric dipole. But first we need to clarify that the calculated and experimental results for the considered models were obtained, assuming that the models use the entire area of the personal phone casing. This assumption is based on uniting an antenna with ground and elimination of a separate antenna and a counterpoise. Components of radio transmitter and receiver can be mounted on the metal antenna, which replace the ground for those elements. Filters, placed on this plate, provide short for direct current and insulation at high frequency.

As previously indicated, if the lengths L_1 and L_2 of the lower and upper arm of the proposed antenna are equal, respectively, to $\lambda/4$ and $\lambda/16$, the active component of the input impedance is close to 30 Ohm, and the reactive component is close to zero. Along the same length of the phone housing one may place at a symmetrical dipole with arm length $\lambda/8$, i.e., smaller twice. Its resistance is

$$R_\Sigma = R_{\Sigma 0}/n^2, \tag{9.19}$$

where $R_{\Sigma 0} \approx 80$ Ohm is the active component of the resonant dipole's input impedance. From (9.19) it follows that in this case $R_\Sigma \approx 20$ Ohm.

It is more substantial that the dipole impedance has a reactive component, which increases the losses in the cable much stronger than the low value of R_Σ. If the length of a dipole arm is $\lambda/8$, its input reactance is equal to wave impedance. Let us take a relatively small wave impedance of 100 Ohm. It is easy to be convinced that in accordance with (2.57) travelling wave ratio (*TWR*) for the dipole with resistance 20 Ohm and reactance 100 Ohm in the cable with $W = 50$ Ohm is equal to 0.08, while *TWR* for a monopole with resistance 30 Ohm and zero reactance in the same cable is 0.6, i.e., 7.5 times greater.

This example shows the obvious advantage of the proposed antenna, resonant dimensions of which fit into the dimensions of the phone length.

This advantage is related with the use of multi-folded structure. It should be emphasized that such a structure can be used not only together with a simple flat monopole, but also with other variants of asymmetric antennas. As can be seen from the given example, such use can dramatically improve the characteristics of existing antennas both by reducing head irradiation, and also by increasing the number of frequency bands for the operation of cell phones. At the same time, increasing the antenna gain and reliability of communication will save an equipment from the necessity to switch the antennas from one frequency to another.

Simulation of the proposed antenna was carried out using the *CST* program, and the results are compared with the results of the planar inverted *F* antenna *(PIFA)*. The results of tuning are given in Fig. 9.14, where the standing wave ratio *(SWR)* of the proposed antenna is shown. The calculation results of reducing field in a near region in the presence of user's head along the perpendicular to the antenna plane are given in Fig. 9.15 (solid curves—in the direction of user, dotted curves—in the opposite direction). The calculation uses the model of the head as part of the program *CST*. The fields are presented at frequencies 0.9 and 1.8 GHz as the functions of distance *S* from the antenna plane.

A full-scale model of the new antenna was fabricated in accordance with results of the calculation. Photo of this model is presented in Fig. 9.16. In Fig. 9.17 the calculated curves and experimental values for the antenna near field at the frequency 0.9 GHz are compared. Fields in the direction of the head and in the opposite direction are denoted by numbers 1 and 2 relatively. As can be seen from the figure, the field of the antenna in the direction of the head is substantially smaller. The experimental values were determined by means of measurement on the phantom model.

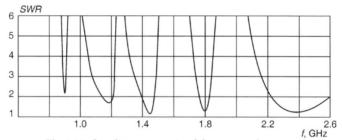

Fig. 9.14: Standing wave ratio of the proposed antenna.

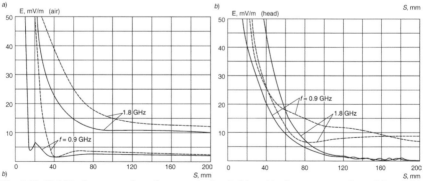

Fig. 9.15: Field in the near region of antenna in the air (a) and in the presence of user's head (b) as a function of distance S from the antenna plane.

The directivity of the offered antenna is presented in Table 9.3. In Table 9.4 the level of *SAR* in the user's head is shown for the proposed antenna and antenna *PIFA* at the frequency 0.9 GHz.

The proposed asymmetrical antenna with the long arm and zero reactance on the operating frequencies has high electrical characteristics and creates in the far region the electromagnetic field, that exceeds significantly the fields of the other antennas of the same dimensions. In particular, the field of the new antenna at the same power is much greater than the field of antenna *PIFA*. The results of researching the new antenna, including simulations and experimental tests, corroborate, that this antenna is promising for use in modern cellular phones. In contrast to known antennas this antenna is multi-frequency, i.e., it can operate on multiple frequencies simultaneously providing high electrical characteristics without switching. Additional advantages of this antenna is decrease of user's head irradiation. It should be emphasized that this method may be used for reducing the irradiation produced by other antennas, in particular by the antenna *PIFA*.

As follows from the contents of this chapter, all these positive qualities are associated with the features of multi-folded antennas.

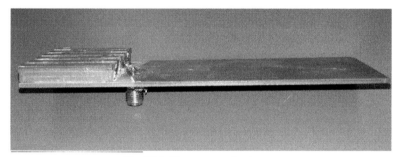

Fig. 9.16: The proposed antenna.

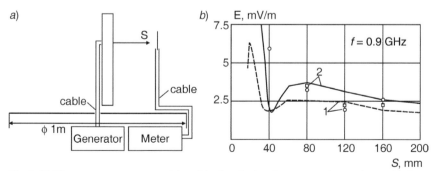

Fig. 9.17: The measurement setup (a) and both calculated curves and experimental values (b) for the near fields at the frequency 0.9 GHz in the direction of the head (1) and in the opposite direction (2).

Table 9.3: Directivity of the proposed antenna.

Frequency, GHz	Directivity, dB
0.9	2.59
1.8	4.48

Table 9.4: SAR of antennas.

Antenna	*f, GHz*	Total	Max local in 10 g	Max local in 1 g	Max point
PIFA	0.9	0.061	3.62	5.8	345.8
Proposed	0.9	0.013	0.762	1.41	21.6

10

Multi-wire and Multi-radiator Antennas

10.1 Multi-wire antenna

In Chapter 8 folded antennas are considered, and in Chapter 9—more complex, multi-folded antennas. Each multi-folded antenna consists of several folded antennas. Accordingly, several simple linear radiators (dipoles, monopoles), when they combine, create more complex structures—multi-wire and multi-radiator antennas. They are also built as systems of parallel wires. This chapter is devoted to them.

Let's start with multi-wire antennas. In order to increase the matching level in a wide frequency range, antennas should have a low wave impedance. For this purpose, the antennas are made of several wires with the same length, arranged along the generatrices of the cylinder (Fig. 10.1a). At the antenna ends the wires usually converge to a common point, forming a cone (Fig. 10.1b). Such radiators are named multi-wire or volumetric, and also cage antennas. If the antenna wires have different lengths (Fig. 10.1c), then the antenna can be considered as a system of several radiators connected to one generator. It is expedient, if it is possible, to chosen the number and lengths of radiators so that in the working frequency band to ensure maximum mutual compensation of radiators reactances and a radiation at small angles to the horizon. Such an antenna is named a multi-radiator one.

When analyzing a multi-wire radiator, it is usually assumed that one can be replaced this radiator by a solid metal tube with an equivalent radius

$$a_e = \sqrt[N]{Na\rho^{N-1}} ,$$ (10.1)

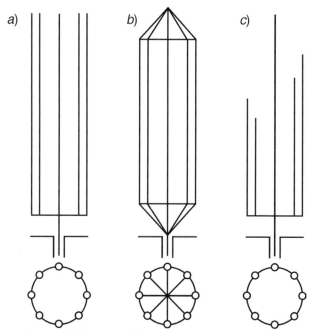

Fig. 10.1: Multi-wire (a, b) and multi-radiator (c) antennas.

where N is the number of wires, a is the wire radius, ρ is the radius of the cylinder along whose generatrices the wires are located. In this case the fields created in the far zone by a multi-wire radiator and by an equivalent metal tube are the same if their lengths are the same.

As can be seen from (10.1), the radius of the equivalent radiator substantially exceeds the radius a of a single wire. Accordingly, the wave impedance W_R decreases and consequently the reactive component X_A of the input antenna impedance decreases.

From the identity of fields in the far zone, generally speaking, the identity of the remaining electrical characteristics, for example, of the input impedances, does not follow. The input impedances of a multi-wire and equivalent radiator were compared with each other in [49]. In this case, the multi-wire radiator (see Fig. 10.1a) was replaced by a structure of N identical linear radiators with the same exciting emf and the same currents.

As calculations had shown, the curves for the input impedance of a multi-wire antenna and the equivalent radiator have the same form but the former are shifted toward larger values of L/λ. As N increases, this shift decreases. The current distribution depends on N and differs from the current distribution along the usual linear radiator ($N = 1$): the current in a separate wire decreases, the wave propagation constant along the antenna grows.

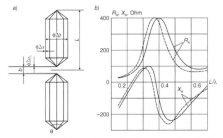

Fig. 10.2: Cage antenna, excited by a two-wire line: a—circuit, b—input impedance.

Similar data for a cage antenna were obtained in [50]. The calculation results for such an antenna excited by a two-wire long line (Fig. 10.2) are given in [51]. The dimensions of an antenna and a long line (in meters) are indicated in the figure: $L = 12.5$, $b = 0.25$, $a = 1.5 \cdot 10^{-3}$, $\theta = 45°$. Each arm of the radiator is formed by eight wires ($N = 8$), a wave impedance of the line is equal to $W_l = 350$ Ohm. The calculation takes into account the spatial coupling between the wires of the radiator and the line and the end effect of the line. In this case, the input impedance of the antenna is understood as the impedance of the load, which, being connected to the line end, creates in the arbitrary cross-section of the line outside the excitation zone the same ratio of the reflected and falling waves of current, as when connecting the antenna. This ratio is not equal to the ratio of the scalar potentials difference between the antenna terminals to the current passing through these terminals. But is equal to the input impedance determined with the help of a measuring instrument connected to the line outside the excitation zone.

The solid line in Fig. 10.2 shows the impedance of a multi-wire cage antenna, the dashed line shows the impedance of an equivalent cylindrical radiator with a radius determined by formula (10.1).

10.2 Multi-radiator antenna

Multi-wire antennas are used, as a rule, in one frequency band, although this band can be quite wide. The multi-radiator antenna can operate as a multi-frequency antenna. In this respect, it is analogous to multi-folded structures. One possible variants of a multi-radiator antenna, which has found practical application, is given in Fig. 10.3a. Its equivalent long line is presented in Fig. 10.3b. The antenna is described in [52].

The antenna consists of a central radiator 1 with a complex impedance load Z_0 and side radiators 2 located around a radiator 1 along the generatrices of a cylinder and connected to the base of the radiator 1. The impedance load is a distinguishing feature of this antenna. It is included to provide a mode of operation close to the running wave mode on the section between the generator and the load. This allows to increase the level of matching

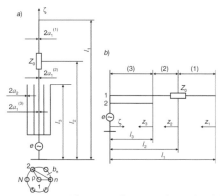

Fig. 10.3: Multi-radiator antenna with a complex impedance load (a) and an equivalent asymmetrical line (b).

and to obtain the desired radiation pattern. As an impedance load, a special power absorber is used. The multi-radiator antenna without a load has high characteristics in the vicinity of series resonances of individual radiators, but in order to obtain such characteristics in a continuous range, its diameter must be significantly increased to reduce the spatial coupling between these radiators. Combining the two principles of antenna construction: the inclusion of a complex load and the use of a multi-radiator structure makes it possible to create an antenna of acceptable dimensions with good characteristics.

To calculate the antenna characteristics, a program based on the method of moments can be used. Another method of calculation, requiring less computer time and having more physical visibility, is described in [53]. In it, for simplicity, it is assumed that the geometrical dimensions of the side radiators are the same, and the radiators are evenly distributed around the circumference of the cylinder, since this variant also gives great possibilities for tuning the antenna to the required frequencies. The result of the solution practically does not depend on the evenness of this placement. But the problem can be solved in the general case, including for different lengths of radiators.

The chosen variant allows to simplify the problem by reducing the asymmetrical line to a two-wire line, and to obtain a solution for the current in explicit form. The first wire of the asymmetric line is the central radiator, and the second wire is formed by a system of $N-1$ side radiators (N is the total number of radiators). Since the wires of the line have different lengths, and one of them has an impedance load, the line should be divided into three sections. Expressions for the current and potential on the section m of the wire n of an asymmetrical line have a form analogous to the expressions (7.8), with $n = 1, 2, m = 1, 2, 3$. The coordinates are presented in Fig. 10.3b.

The wire numbers are given at the left end of the wires, the section numbers are indicated in round brackets in the middle of each section.

The boundary conditions for the two-wire asymmetric line shown in Fig. 10.3b, are written as

$$i_1^{(1)}\Big|_{z_1=0}=i_2^{(3)}\Big|_{z_3=0},i_1^{(1)}\Big|_{z_1=l_1-l_2}=i_1^{(2)}\Big|_{z_2=0},i_1^{2}\Big|_{z_2=l_2-l_3}=i_1^{(3)}\Big|_{z_3=0},$$

$$u_1^{(1)}\Big|_{z_1=l_1-l_2}=u_1^{(2)}-i_1^{(2)}Z_0\Big|_{z_2=0},u_1^{(2)}\Big|_{z_2=l_2-l_3}=u_1^{(3)}\Big|_{z_3=0},u_1^{(3)}\Big|_{z_3=l_3}=u_2^{(3)}\Big|_{z_3=l_3}=e.$$

(10.2)

These conditions mean the absence of current at the free ends of the wires and the continuity of current and potential along each wire, except for the point where the load is set, in which the potential jump takes place. Substituting (7.8) into (10.2) and solving the resulting system of equations, we find:

$$I_1^{(1)}=I_2^{(3)}=0,I_1^{(2)}=j\frac{U_1^{(1)}}{\rho_{11}^{(1)}}\sin k\left(l_1-l_2\right),I_1^{(3)}=j\frac{U_1^{(1)}\sin k\left(l_1-l_2\right)\sin k\left(l_{1e}-l_3\right)}{\rho_{11}^{(1)}\sin k\left(l_{1e}-l_2\right)},$$

$$U_1^{(2)}=\frac{U_1^{(1)}\rho_{11}^{(2)}}{\rho_{11}^{(1)}}\sin k\left(l_1-l_2\right)\cot k\left(l_{1e}-l_3\right),U_1^{(3)}=\frac{U_1^{(1)}\rho_{11}^{(2)}\sin k\left(l_1-l_2\right)\cos k\left(l_{1e}-l_3\right)}{\rho_{11}^{(1)}\sin k\left(l_{1e}-l_2\right)},$$

$$U_2^{(3)}=\frac{U_1^{(1)}\rho_{11}^{(2)}\sin k\left(l_1-l_2\right)\cos k\left(l_{1e}-l_3\right)}{\rho_{11}^{(1)}\sin k\left(l_{1e}-l_2\right)}\left[1-\frac{\rho_{11}^{(3)}-\rho_{12}^{(3)}}{\rho_{11}^2}\tan k\left(l_{1e}-l_3\right)\tan kl_3\right],$$

$$U_1^{(1)}=-e\rho_{11}^{(1)}\sin k\left(l_{1e}-l_2\right)\left\{\rho_{11}^{(3)}\sin k\left(l_1-l_2\right)\sin k\left(l_{1e}-l_3\right)\sin kl_3\left[1-\frac{\rho_{11}^{(2)}}{\rho_{11}^{(3)}}\cot k\left(l_{1e}-l_3\right)\cot kl_3\right]\right\}.$$

Here it is taken into account that $W_{11}^{(1)}=\rho_{11}^{(1)},W_{11}^{(2)}=\rho_{11}^{(2)}$, and l_{1e} is a complex quantity determined from expression $Z_0-j\rho_{11}^{(1)}\cot k\left(l_1-l_2\right)=-j\rho_{11}^{(2)}\cot k\left(l_{1e}-l_2\right)$.

The total current along the antenna is a function of the coordinate $\zeta=l_m-z_m$ (see Fig. 10.3b).

$$J(\zeta)=\sum_{n=1}^{M}i_n^{(m)}=\begin{cases}j\dfrac{U_1^{(1)}}{\rho_1^{(1)}}\sin k\left(l_1-|\zeta|\right),l_2\le|\zeta|\le l_1,\\[3mm]j\dfrac{U_1^{(1)}}{\rho_1^{(1)}}D_1\dfrac{\sin k\left(l_{1e}-|\zeta|\right)}{\sin k\left(l_{1e}-l_3\right)},l_3\le|\zeta|\le l_2,\\[3mm]j\dfrac{U_1^{(1)}}{\rho_1^{(1)}}\left[D_1\cos k\left(l_3-|\zeta|\right)+D_2\sin k\left(l_3-|\zeta|\right)\right],0\le|\zeta|\le l_3,\end{cases}$$

(10.3)

where

$$D_1 = \frac{\sin k(l_1 - l_2)\sin k(l_{1e} - l_3)}{\sin k(l_{1e} - l_2)},$$

$$D_2 = \rho_{11}^{(2)} \frac{\sin k(l_1 - l_2)\cos k(l_{1e} - l_3)}{\sin k(l_{1e} - l_2)} \left\{ \left[\frac{1}{W_{11}^{(3)}} - \frac{1}{W_{12}^{(3)}} \right] + \left[\frac{1}{W_{22}^{(3)}} - \frac{1}{W_{12}^{(3)}} \right] \left[1 - \frac{\rho_{11}^{(3)} - \rho_{12}^{(3)}}{\rho_{11}^{(3)}} \tan k(l_{1e} - l_3)\tan kl_3 \right] \right\}.$$

Input impedance of an asymmetrical line

$$Z_1 = e/J(0). \tag{10.4}$$

This expression, taking into account (10.3), permits to approximately determine a reactive impedance of a multi-radiator antenna—similar to how an expression for an input impedance of an equivalent two-wire long line permits to approximately determine a reactive impedance of a linear radiator. More precisely, the antenna impedance can be found by considering a linear radiator, the current along which is equal to the total current along the multi-radiator antenna.

In accordance with the second formulation of the method of induced emf, the impedance of considering antenna is

$$Z_A = -\frac{1}{J^2(0)} \left[\int_0^{l_1} E_\zeta J(\zeta) d\zeta - Z_0 J^2(l_2) \right], \tag{10.5}$$

where E_ζ is the tangential component of an electric field created on the radiator surface by the current $J(\zeta)$ along its axis, which is find from (10.3). The free term in the square brackets of expression (10.5) is the power that is lost in a complex load Z_0. In the vicinity of parallel resonance, this expression gives a large error, and it should be replaced by

$$Z_A = \frac{e}{2J(0) + \frac{1}{e}\int_0^{l_1} E_\zeta(J)d\zeta - \frac{1}{e}Z_0 J^2(l_2)}. \tag{10.6}$$

A field E_ζ is calculated by the usual way. A function $J(\zeta)$ is continuous on the whole interval of integration and is a sinusoid at each point. However, the function $dJ/d\zeta$ suffers a break at the sections boundaries, i.e., it is necessary to take into account the difference in the derivatives on both sides of the discontinuity.

The antenna radiation resistance is

$$R_\Sigma = R_A - R_{load},\qquad(10.7)$$

where R_A is the active component of the input impedance, obtained with the help of (10.5) or (10.6), and R_{load} is the loss resistance in the active load $Z_0 = R_0$, referred to the antenna input, and

$$R_{load} = \mathrm{Re}\frac{J^2(l_2)R_0}{J^2(0)}.\qquad(10.8)$$

Indeed, in accordance with the theorem on oscillating power $P = P_1 + P_2$, where $P = J^2(0)(R_A + jX_A)$ is an oscillating power given by the source, $P_1 = J^2(0)(R_\Sigma + jX_\Sigma)$ is an oscillating power passing through the surface of the antenna, $P_2 = J^2(l_2)R_0$ is the oscillating power in the load, i.e.,

$$R_A = \mathrm{Re}\frac{P_1 + P_2}{J^2(0)} = \mathrm{Re}\frac{P_1}{J^2(0)} + \mathrm{Re}\frac{J^2(l_2)R_0}{J^2(0)}.$$

These expressions follow from the equalities (10.7) and (10.8).

The expression (10.3) for the total current of the multi-radiator antenna includes wave impedances $\rho_{ns}^{(m)}$ and $W_{ns}^{(m)}$ whose values are determined by the potential coefficients. For an asymmetrical line of two wires with different lengths, divided into three sections:

$$\rho_{ns}^{(m)} = \frac{\alpha_{ns}^{(m)}}{c},\frac{1}{W_{11}^{(3)}} = \frac{\rho_{22}^{(3)}}{\Delta},\frac{1}{W_{12}^{(3)}} = \frac{\rho_{12}^{(3)}}{\Delta},\frac{1}{W_{22}^{(3)}} = \frac{\rho_{11}^{(3)}}{\Delta},\qquad(10.9)$$

where $\Delta = \rho_{11}^{(3)}\rho_{22}^{(3)} - [\rho_{12}^{(3)}]^2$. Potential coefficients are calculated by the method of average potentials, based on the actual location of the antenna wires. Taking into account the mirror image, we obtain

$$\alpha_{ns}^{(m)} = \frac{1}{2\pi\varepsilon}\left\{\alpha\left[l_m - l_{m+1}, 0, b_{ns}^{(m)}\right] - \alpha\left[l_m - l_{m+1}, l_m + l_{m+1}, b_{ns}^{(m)}\right]\right\},\qquad(10.10)$$

where $a(L,l,b)$—the mutual potential coefficient for two parallel wires shown in Fig. 7.2. Its magnitude is determined by the expression (7.10). The parameter $b_{ns}^{(m)}$ in expression (10.10) is the distance between the axes of wires n and s in the section m.

For the first wire of an asymmetrical long line

$$\alpha_{11}^{(m)} = \frac{1}{2\pi\varepsilon}\left\{\alpha\left[l_m - l_{m+1}, 0, a_1^{(m)}\right] - \alpha\left[l_m - l_{m+1}, l_m + l_{m+1}, a_1^{(m)}\right]\right\}, \quad (10.11)$$

where $a_1^{(m)}$ is radius of the central radiator on section m. The self-potential coefficient of the second wire of the long line is equal to its potential at $Q_1 = 0$, $Q_2 = 1$ (Q_i is the charge per unit length of the i wire), and the charge Q_2 is evenly distributed among the side radiators due to symmetry: $q_n = Q/(N-1)$. Therefore

$$\alpha_{22}^{(3)} = \sum_{n=2}^{N}\alpha_{2n}\frac{q_n}{Q_2} = \frac{1}{N-1}\sum_{n=2}^{N}\alpha_{2n}. \quad (10.12)$$

where $\alpha_{2n} = \frac{1}{2\pi\varepsilon}\left[\alpha\left(l_3, 0, b_n\right) - \alpha\left(l_3, l_3, b_n\right)\right]$ is the mutual potential coefficient between the second and n radiators, and

$$b_n = \begin{cases} 2\rho\sin\dfrac{\pi(n-2)}{N-1}, 3 \le n \le N, \\ a_2, n = 2, \end{cases}$$

where ρ is the radius of a cylinder, along the generatrices of which side radiators are located, a_2 is the radius of the side radiator.

The mutual potential coefficient between the first and second wire of the long line is equal to the potential of the first wire at $Q_1 = 0$, $Q_2 = 1$, i.e.,

$$\alpha_{12}^{(3)} = \sum_{n=2}^{N}\alpha_{1n}\frac{q_n}{Q_2} = \alpha_{1n}, \quad (10.13)$$

where $\alpha_{1n} = \frac{1}{2\pi\varepsilon}\left[\alpha\left(l_3, 0, \rho\right) - \alpha\left(l_3, l_3, \rho\right)\right]$ is the mutual potential coefficient between the first and n radiator.

Figure 10.4 presents the characteristics of the multi-radiator antenna shown in Fig. 10.3b. It has six side radiators. Dimensions (in meters) are $l_1 = 10, l_2 = 7, l_3 = 6.5, a_1^{(1)} = 0.007, a_1^{(2)} = a_1^{(3)} = 0.02, a_2 = 0.01, \rho = 0.15$.

Load Z_0 represents a parallel connection of a resistor with a resistance $R = 200$ Ohm and a coil with an inductance $\Lambda = 14\mu H$. Together with the calculated curves in Fig. 10.4 are given experimental values quite good coinciding with them.

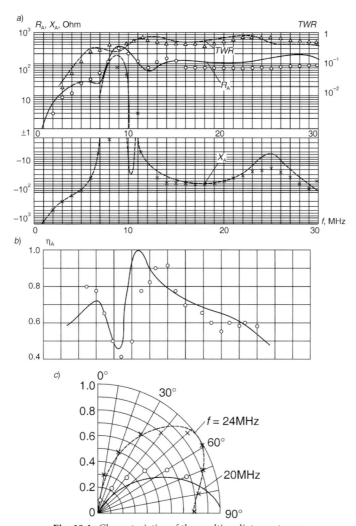

Fig. 10.4: Characteristics of the multi-radiator antenna.

11

Multi-tiered and Log-periodic Coaxial Antennas

11.1 Multi-tiered antenna

Wide-band and multi-frequency antennas perform a common task—increase the number of points in the frequency range with a high level of signal transmission. But the solution of this problem is based on different, more precisely, opposite principles. In a wide-band antenna, one complex device is used, designed to operate at all frequencies. A multi-frequency antenna is a set of simple identical devices operating at one frequency or in a narrow frequency band. This principle cannot be considered rigid. It is often mitigated when several complex devices are used in a wide-band antenna, and in the multifrequency antenna the identical devices have a complex structure. This is not hard to be seen from the examples of antennas considered in the book.

Similar simple devices of different and identical sizes, for example, linear radiators, folded radiators, simple self-complementary antennas are assembled in multifrequency antennas: multi-radiator, multi-folded, with rotation symmetry. From this list it can be seen that the elements of the multifrequency antenna are selected mainly based on the need to ensure a high level of matching. In this chapter antennas are considered, the elements of which provide the required directional patterns in a vertical plane for different parts of the range.

Vertical linear radiators, for example whip antennas, are widely used for a short-wave radio communication of mobile objects. They take up little space and do not interfere with overview. Their disadvantage is that the maximum radiation is directed horizontally only when the electrical length of the antenna is small. With the growth of an electrical length the horizontal signal decreases. If the antenna height L is greater than 0.7λ, a main lobe of a radiation pattern in a vertical plane separates from a ground and the radiation in the horizontal direction drops sharply. Changing of the antenna height, for example by means of a telescopic construction, allows to improve the antenna characteristics. But such mechanical tuning consumes a time and complicates the antenna design. Folded structures allow to create an antenna, in which the length of the radiating element is changed without changing the geometric dimensions of the device [54].

In the simplest version, the antenna is designed for operation in two frequency bands (similar to the telescopic antenna, which may have two geometrical heights during operation). The antenna is made as two-tiered and consists of two radiators, which are located one above the other and connected together. The upper radiator is linear, the lower radiator is folded (Fig. 11.1). Excitatory emfs e_1 and e_2 are included in both wires of the folded radiator and may change in amplitude and phase by means of tuning circuit.

If the antenna operates in the first frequency band, emfs have the same phase, creating in both wires of folded radiator the currents in one direction. These currents become by the current of the single unified linear radiator. As a result, the current is created along the all antenna, and the height of radiating section is equal to the total height of the antenna.

If the antenna operates in the second frequency band, the current is created only in the wires of the folded radiator. With this aim excited emfs are included in anti-phase, and their magnitudes are chosen so that the potential at the point of connecting the linear and folded radiators was zero. This eliminates the immediate excitation of the linear radiator. The upper section of the antenna, consisting of one wire, can be excited by electromagnetic fields of the currents flowing along the wires of the folded radiator. However, if the length of the upper segment is far from the resonance (it is not a multiple of $\lambda/2$), the current along this segment is small, i.e., the height of the radiating section of the antenna is equal to the height of the folded radiator.

Changing the radiator height allows to provide the operation in two frequency bands. If the frequency ratio in each band is equal to k_f, the total frequency ratio is equal to k_f^2. The height of the folded radiator is chosen to be equal to

$$l = L/k_f. \tag{11.1}$$

The circuit of considered antenna can be generalized for using when the number of frequency bands is N (instead two bands). For this, the number of antenna tiers should be increased to N. The two upper tiers are similar with the described version. The lower ends of each wire of the folded radiator connect with the upper points of the folded radiator of the next

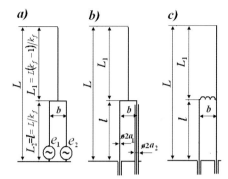

Fig. 11.1: Two-tiered antenna: a—total circuit, b—with coaxial cable, c—with reactive load

(third) tier, etc. In the case of N tiers overall frequency overlap is equal to k_f^N. The heights of the lower tier and the rest tiers are given by expressions

$$L_n = \begin{cases} L/k_f^{n-1}, n = N, \\ L(k_f - 1)/k_f^n, n \neq N, \end{cases} \tag{11.2}$$

where n is the tier number, counting from the top.

Multi-tiered antenna creates a new prospect for the development of a wide-range antenna. In this direction the most significant results were obtained earlier by means of including concentrated capacitive loads and optimization of these loads (see Chapters 3 and 4). Calculations show (see Fig. 3.5) that the capacitive loads allow extending the range in the direction of high frequencies with a sufficiently high level of matching, ensuring the frequency ratio of the order of 10. Using of the multi-tiered structure and the capacitive loads in the wires of each tier allows to ensure in a wide range the high level of matching and the required directional pattern.

Let us return to the two-tiered variant of the antenna, more precisely to its excitation in anti-phase mode. The antenna will radiate, if to provide asymmetry in the folded radiator, i.e., it is necessary to obtain different amplitudes of the currents in the left and the right branches of the radiator. With this aim one of the wires must be accomplished in the form of a coaxial cable (see Fig. 11.1b), i.e., generator must be included not in the lower, but in the upper point of the wire. Another way of creating an asymmetry is inclusion of a reactive load in one of the wires (see Fig. 11.1c).

In the presence of asymmetry not only anti-phase, but also in-phase currents will flow in the branches of the folded radiator. These currents are caused by the presence of the ground (see the last paragraph of Section 8.3 and Fig. 8.11). Namely these currents create radiation. However, for the sake of simplicity we shall conventionally name by anti-phase mode the mode of antenna operation, when the potential at the point of connecting linear and folded radiators is zero.

In order to analyze the two-tiered antenna, we shall apply the theory of electrically coupled long lines, described in Section 7.1. This theory allows to find the currents and the potentials along each wire of the line and emf of generators providing the required operation mode. In this case, the equivalent line (Fig. 11.2) is considered in the general form—with two generators in one of the branches and two complex loads. The set of equations (7.3) for the three wires in this case takes the form:

$$i_1 = I_1 \cos kz_1 + j[U_1/W_{11} - U_2/W_{12}]\sin kz_1, \quad u_1 = U_1 \cos kz_1 + j(\rho_{11} I_1 + \rho_{12} I_2)\sin kz_1,$$

$$i_2 = I_2 \cos kz_1 + j[U_2/W_{22} - U_1/W_{12}]\sin kz_1, \quad u_2 = U_2 \cos kz_1 + j(\rho_{12} I_1 + \rho_{22} I_2)\sin kz_1,$$

$$i_3 = I_3 \cos kz_3 + j(U_3/W_{33}) \sin kz_3, \quad u_3 = U_3 \cos kz_3 + j\rho_{33} I_3 \sin kz_3. \quad (11.3)$$

The boundary conditions for the currents and the potentials are

$$i_3\big|_{z_3 = 0} = 0, \ i_1 + i_2\big|_{z_1 = 0} = i_3\big|_{z_3 = L - l}, \ u_1\big|_{z_1 = 1} = e_1, \ u_2\big|_{z_1 = l} = e_2,$$

$$u_1 - Z_1 i_1\big|_{z_1 = 0} = u_2 - Z_2 i_2\big|_{z_1 = 0} + e_3 = u_3\big|_{z_3 = L - l}. \quad (11.4)$$

Equalities (11.3) and the boundary conditions (11.4) are the set of equations with six unknown magnitudes U_i, I_i (i = 1, 2, 3). Substituting (11.3) into (11.4), we obtain:

$$I_3 = 0, \quad I_2 = -I_1 + j(U_3/W_{33})\sin k\,(L - l), \quad U_1 = Z_1 I_1 + U_3 \cos k\,(L - l),$$

$$U_2 = -e_3 - Z_2 I_1 + U_3 \cos k\,(L - l)[1 + j(Z_{22}/W_{33}) \tan k\,(L - l)],$$

$$e_1 = I_1 \cos kL\left[Z_1 + j\left(\rho_{11} - \rho_{12}\right)\tan kL\right] +$$
$$U_3 \cos kL \cos k\,(L - l)\left[1 - \left(\rho_{12}/W_{33}\right)\tan kL \tan k\,(L - l)\right],$$

$$e_2 = -e_3 \cos kL - I_1 \cos kL\left[Z_2 + j\left(\rho_{22} - \rho_{12}\right)\tan kL\right] +$$
$$U_3 \cos kL \cos k\,(L - l)\left[1 - \frac{\rho_{22} - jZ_2}{W_{33}} \tan kL \tan k\,(L - l)\right]. \quad (11.5)$$

The rest two formulas permit to express the magnitudes I_1 and U_3 through emfs of the generators. But the corresponding expressions are cumbersome. They are not used for further analysis and not presented here.

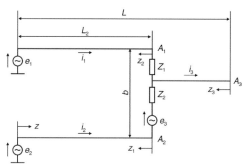

Fig. 11.2: Equivalent asymmetric line.

The previous four equalities after substituting into (11.3) allow to find the current distribution along each wire as a function of magnitudes I_1 and U_3:

$$i_1(z) = I_1 \cos k(L-z)\,[1 + j(Z_1/W_{11} + Z_2/W_{12})\tan k(l-z)] + j(e_3/W_{12})\sin k(l-z) +$$

$$+ U_3[j(1/W_{11} - 1/W_{12})\cos k(L-l) + Z_2 \sin k(L-l)/(W_{12}\,W_{33})]\sin k\,(L-z),$$

$$i_2(z) = -I_1 \cos k(L-z)\left[1 + j\left(\frac{Z_1}{W_{12}} + \frac{Z_2}{W_{22}}\right)\tan k(L-z)\right] - j\frac{e_3}{W_{22}}\sin k(L-z) +$$

$$U_3\left[j\left(\frac{1}{W_{22}} - \frac{1}{W_{12}}\right)\cos k(L-l) - \frac{Z_2}{W_{22}W_{33}}\sin k(L-l) + j\frac{1}{W_{33}}\sin k(L-l)\cot k(L-z)\right]\sin k(L-z).$$

Further we shall consider the specific versions of antennas as realization in particular cases of general equivalent circuit. The circuit of two-tiered antenna with the coaxial cable is shown in Fig. 11.1b. Here a few of elements of the overall circuit is absent, i.e.,

$$Z_1 = Z_2 = e_2 = 0. \tag{11.6}$$

In the anti-phase mode in accordance with the boundary condition $u_3|_{z_3=L-l} = 0$ we find that $U_3 = 0$. Then from (11.5) we obtain emf of generators

$$e_1 = j(\rho_{11} - \rho_{12})I_1 \sin kL,$$

$$e_3 = -j(\rho_{22} - \rho_{12})I_1 \tan kL = -je_1(\rho_{22} - \rho_{12})\sec kL/(\rho_{11} - \rho_{12}). \tag{11.7}$$

It is necessary to emphasize that the relationship between emf of two generators is obligatory condition for providing anti-phase mode in the antenna. The currents along the antenna wires in this mode according to (11.6) are

$$i_1(z) = I_1 \cos k(l-z) + \frac{\rho_{22} - \rho_{12}}{W_{12}}I_1 \tan kl \sin k(l-z),$$

$$i_2(z) = -I_1 \cos k(l-z) - \frac{\rho_{22} - \rho_{12}}{W_{12}}I_1 \tan kl \sin k(l-z), \quad i_3(z) = 0. \tag{11.8}$$

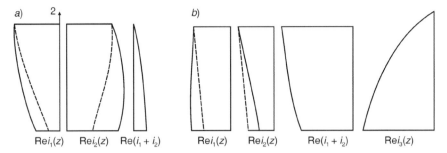

a) b)

$\mathrm{Re}i_1(z)$ $\mathrm{Re}i_2(z)$ $\mathrm{Re}(i_1 + i_2)$ $\mathrm{Re}i_1(z)$ $\mathrm{Re}i_2(z)$ $\mathrm{Re}(i_1 + i_2)$ $\mathrm{Re}i_3(z)$

Fig. 11.3: Currents in the two-tiered antenna with a coaxial cable in anti-phase (a) and in-phase (b) modes.

The expressions (11.8) confirm that the currents along the first and the second wires contain in-phase and anti-phase components. The total antenna current (sum of currents) varies along the antenna similarly to the current along the linear radiator of the length l

$$i_1(z) + i_2(z) = -\frac{(\rho_{22} - \rho_{12})(\rho_{11} - \rho_{12})}{\rho_{11}\rho_{22} - \rho_{12}^2} I_1 \tan kl \sin k(l - z). \qquad (11.9)$$

Here it is taken into account that (see Section 7.1)

$$\frac{1}{W_{22}} = \frac{\rho_{11}}{\rho_{11}\rho_{22} - \rho_{12}^2}, \frac{1}{W_{12}} = \frac{\rho_{12}}{\rho_{11}\rho_{22} - \rho_{12}^2}.$$

The current distribution along the antenna wires in the anti-phase mode is shown in Fig. 11.3a. The currents in the first and second wires are denoted by the numerals 1 and 2, the total currents—by the sum of $1 + 2$. Impedance on the output of each generator, which excites asymmetric long line, is equal to

$$jX_{A1} = \frac{e_1}{i_1(0)} = j(\rho_{11} - \rho_{12})\tan kl \left[1 + \frac{\rho_{22} - \rho_{12}}{W_{12}} \tan^2 kL\right]^{-1},$$

$$jX_{A3} = \frac{e_3}{i_2(l)} = j(\rho_{22} - \rho_{12})\tan kl. \qquad (11.10)$$

These expressions allow to determine approximately the reactive component of the loading impedance of each generator (similarly to the formula for the input impedance of an equivalent two-wire long line, which allows determining approximately the reactive component of the input impedance of a linear antenna).

From the viewpoint of radiation, as seen from (11.8) and Fig. 11.3a, the antenna in anti-phase mode consists of two parallel radiators of the height l with in-phase currents in the base:

$$i_1^{(i)} = \frac{\rho_{22} - \rho_{12}}{W_{12}} I_1 \tan kl \sin kl, \quad i_2^{(i)} = -\frac{\rho_{22} - \rho_{12}}{W_{22}} I_1 \tan kl \sin kl.$$

The radiation resistance of each radiator consists of the self-resistance and the mutual resistance multiplied by the ratio of the currents. In particular, for the first radiator we write:

$$R_{A1} = R_{11} + R_{12} \frac{i_2^{(i)}(0)}{i_1^{(i)}(0)}, \tag{11.11}$$

where R_{11} is self-radiation resistance, R_{12} is the mutual radiation resistance, and $R_{12} \approx R_{11}$, since the radiator heights are the same and the distance between the radiators is small in comparison with the wave length. Thus,

$$R_{A1} = R_{11}(1 - W_{12}/W_{22}), \quad R_{A2} = R_{11}(1 - W_{22}/W_{12}). \tag{11.12}$$

In the anti-phase mode the electric field strength in the far region and the directional pattern coincide with the similar characteristics of the conventional linear radiator of a height l. Expressions (11.8)–(11.12) are sufficiently simple and allow determining the influence of antenna geometric dimensions upon the current magnitude in each wire and upon the electrical characteristics of the radiator. More precisely, the input impedance of the antenna can be calculated by using an algorithm of calculation, based on the integral equation for the current, the Moment Method and the systems of piecewise sinusoidal basis functions.

Let's move on to an analysis of the in-phase mode. For the implementation of this mode one must ensure equality of potentials in both branches of the folded radiator, i.e.,

$$u_1(z) = u_2(z). \tag{11.13}$$

Applying this condition to the set of equations (11.3), we find:

$$e_1 = jI_1 \frac{\rho_{11}\rho_{22} - \rho_{12}^2}{\rho_{22} - \rho_{12}} \sin kl - jI_1 W_{33} \frac{\rho_{11} + \rho_{22} - 2\rho_{12}}{\rho_{22} - \rho_{12}} \cot k(L-l), \quad e_3 = e_1 \sec l. \tag{11.14}$$

Currents along the wires consist in this case of the in-phase components only:

$$i_1(z) + i_2(z) = I_1 W_{33} \frac{\rho_{11} + \rho_{22} - 2\rho_{12}}{\rho_{11}\rho_{22} - \rho_{12}^2} \cot k(L-l) \sin k(l-z) + I_1 \frac{\rho_{11} - \rho_{12}}{\rho_{22} - \rho_{12}} \tan kl \sin k(l-z) +$$

$$+ I_1 \left(1 + \frac{\rho_{11} - \rho_{12}}{\rho_{22} - \rho_{12}}\right) \cos k(l-z), \quad i_3(z) = I_1 \left(1 + \frac{\rho_{11} - \rho_{12}}{\rho_{22} - \rho_{12}}\right) \frac{\sin k(L-z)}{\sin k(L-l)}. \tag{11.15}$$

The current distribution along the wires is shown in Fig. 11.3b. Impedances on the output of each generator, which excites asymmetric long line, are

$$jX_{A1} = \frac{e_1}{i_1(0)} = j\frac{\rho_{11}\rho_{22} - \rho_{12}^2}{\rho_{22} - \rho_{12}} \tan kl \frac{1 - W_{33} \dfrac{\rho_{11} + \rho_{22} - 2\rho_{12}}{\rho_{11}\rho_{22} - \rho_{12}^2} \cot kl \cot k(L-l)}{1 + \dfrac{W_{33}}{W_{11}} \dfrac{\rho_{11} + \rho_{22} - 2\rho_{12}}{\rho_{22} - \rho_{12}} \tan kl \cot k(L-l) - \dfrac{\rho_{12}}{\rho_{22} - \rho_{12}} \tan^2 kl},$$

$$jX_{A3} = \frac{e_3}{i_2(l)} = j\frac{\rho_{11}\rho_{22} - \rho_{12}^2}{\rho_{22} - \rho_{12}} \tan kl \left[1 - W_{33} \frac{\rho_{11} + \rho_{22} - 2\rho_{12}}{\rho_{11}\rho_{22} - \rho_{12}^2} \cot kl \cot k(L-l) \right]. \quad (11.16)$$

The radiation resistance is calculated according to formulas similar to (11.11). But R_{11} is the resistance of the radiator of height L, the current along which is determined by (11.15). Since the derivative of the current has discontinuity on the border of sections, the calculation should take into account the break of the current derivative. Correspondingly it is necessary to replace the known expression for E_z by the equality of type

$$E_z(J) = -j\frac{15}{k}\left\{ \frac{2\exp(-jkR_0)}{R_0} \frac{dJ(0)}{dz} - \left[\frac{\exp(-jkR_1)}{R_1} + \frac{\exp(-jkR_1)}{R_1} \right] \frac{dJ(l_1)}{dz} + \right.$$
$$\left. + \sum_{m=1}^{M} \left[\frac{\exp(-jkR_{m1})}{R_{m1}} + \frac{\exp(-jkR_{m2})}{R_{m2}} \right] \left[\frac{dJ(l_m + 0)}{dz} - \frac{dJ(l_m - 0)}{dz} \right] \right\}, \quad (11.17)$$

where R_{m1} and R_{m2} are the distances from observation point to the sections' borders in the upper and the lower arms of the radiator, M is number of borders, $\dfrac{dJ(l_m + 0)}{dz}$ and $\dfrac{dJ(l_m - 0)}{dz}$ are values of the current derivatives from the left and the right of point $z = l_m$.

It should be noted that for the same diameters of the antenna wires the formulas become far simpler. As an example of the two-tiered antenna with coaxial cable we consider the antenna with dimensions (in meters): $L = 1.0$, $l = 0.39$, $b = 0.037$, $a_1 = 0.002$, $a_2 = 0.025$. Here, b is distance between the axes of the wires of the folded radiators, a_1 and a_2 are the radii of the wires (see Fig. 11.1b). Calculation of the antenna characteristics is made by means of the moment method.

Figure 11.4 shows the calculated curves for the directional patterns in the vertical plane—in the in-phase (a) and anti-phase (b) modes. Model of the antenna was made in full size. The results of the experimental verification are given for frequencies 150 and 300 MHz. As can be seen from the figures, the coincidence of the calculation and the experiment is quite satisfactory. The high level of radiation in the direction perpendicular to the axis of the radiator (along the ground) is provided in the double frequency range. However, in the anti-phase mode when the length of the third wire (of the upper segment of the antenna) is a multiple of half the wavelength, i.e., at frequencies 245 and 490 MHz, the main lobe of the directional pattern is located under a great angle to the horizontal. Here the current along the third wire is too large. The dimensions of the antenna must be chosen such that the resonance frequencies were lying outside the operating range.

Figure 11.5 shows the current distribution along the antenna wires in the anti-phase mode, including the current distribution along the left wire and the connecting bridge between the wires of the folded radiator, and also the current distribution along the right wire and the upper (third) wire. These distributions of currents are the graphic illustration of the processes in the anti-phase mode of the two-tiered antenna. The results are shown at four frequencies, including frequencies 245 and 490 MHz, where the main lobe is located at a large angle to the horizon, since the current of the third (upper) wire is too great.

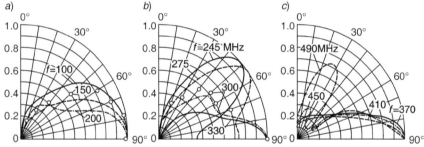

Fig. 11.4: Directional patterns of antenna in the vertical plane in the in-phase (a) and anti-phase (b) modes.

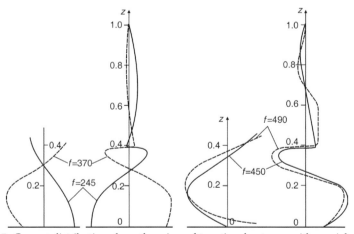

Fig. 11.5: Current distribution along the wires of two-tiered antenna with coaxial cable at four frequencies (in anti-phase mode).

The circuit of two-tier antenna with a reactive load is shown in Fig. 11.1c. Here a few of elements of the overall circuit is absent also, i.e., $Z_2 = e_3 = 0$. Let us assume that $\rho_{11} = \rho_{22}$. In the anti-phase mode the boundary condition $u_3|_{z_3 = L - l} = 0$ should be executed. This means that the emfs of generators are connected by the relationship

$$e_2 = -e_1 + Z_1 I_1 \cos kl. \tag{11.18}$$

Currents along the antenna wires in this mode

$$i_1(z) + i_2(z) = jZ_1 I_1 \sin k(L - z) / (\rho_{11} + \rho_{12}), \ i_3(z) = 0. \tag{11.19}$$

Reactive components of generators' load are

$$jX_{A1} = \frac{Z_1 + j(\rho_{11} - \rho_{12})\tan kl}{1 + j(Z_1 / W_{11})\tan kl}, \ jX_{A2} = \frac{j(\rho_{11} - \rho_{12})\tan kl}{1 + j(Z_1 / W_{12})\tan kl}, \tag{11.20}$$

resistances are calculated in accordance with (11.11), and

$$i_2^{(i)}(0)/i_1^{(i)}(0) = W_{11}/W_{12}.$$

The in-phase mode has the salient features. Since the load is connected in the upper section of the folded radiator, it is impossible to produce the equality of voltages in both its wires. Let us assume

$$u_1(z) - u_2(z) = U \cos k(l - z).$$

For executing this condition it is necessary that

$$e_2 = e_1 - Z_1 I_1 \cos kl. \tag{11.21}$$

The currents along the wires are calculated in accordance with expressions

$$i_1(z) + i_2(z) = 2I_1 \cos k(l - z) + \frac{4I_1 W_{33}}{\rho_{11} + \rho_{12}}\cot k(L - l)\sin k(l - z) + j\frac{Z_1 I_1}{\rho_{11} + \rho_{12}}\sin k(l - z), i_3(z) = 2I_1 \frac{\sin k(L - z)}{\sin k(L - l)}. \tag{11.22}$$

Reactive impedances of generators' load are

$$jX_{A1} = \frac{Z_1 + j(\rho_{11} + \rho_{12})\tan kl - 2jW_{33}\cot k(L - l)}{1 + j(Z_1 / W_{11})\tan kl + \dfrac{2W_{33}}{\rho_{11} + \rho_{12}}\tan kl \cot k(L - l)},$$

$$jX_{A2} = \frac{j(\rho_{11} + \rho_{12})\tan kl - 2jW_{33}\cot k(L - l)}{1 - j\dfrac{Z_1}{W_{12}}\tan kl + \dfrac{2W_{33}}{\rho_{11} + \rho_{12}}\tan kl \cot k(L - l)}. \tag{11.23}$$

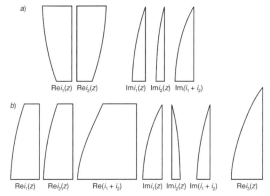

Fig. 11.6: Currents in the antenna with the reactive load in anti-phase (a) and in-phase (b) modes.

The current distribution along the wires of the antenna with the reactive load in the in-phase and anti-phase modes is shown in Fig. 11.6. The in-phase currents are designated by symbol (i), the anti-phase currents are designated by symbol (a).

Summarizing the results presented in this section, one should make the following conclusion. The principle of changing electrical height of the radiator without changing its geometric dimensions, realized in the circuit of two-tiered antenna, is very promising, quite efficient and requires careful study to implement it in real structures.

11.2 Log-periodic coaxial antennas

Log-periodic dipole antenna (LPDA) is a widely used variant of multi-frequency, wide-range and directional antennas. It consists of a large number of linear radiators located along the distribution line. Their lengths and distances from the point of an antenna excitation correspond to the law of geometric progression. The linear radiators sequentially radiate at the frequencies of their first (series) resonances. Additional important features of the antenna are using the complementarity principle and the ability to automatically "cut-off" the current, which turns the antenna into a frequency-independent one. *LPDA* provides directional radiation along its longitudinal axis and preserves shape of the directional pattern over a wide frequency range. It has also constant input impedance. The creation of *LPDA* represents a serious step forward in the development of wide-range directional antennas [55–58].

In accordance with the principle of electrodynamics similarity any radiator has the same electrical characteristics at different frequencies, if its geometric dimensions vary with frequency in proportion to the wavelength (in the first approximation the requirement about corresponding change of the material conductivity may be neglected). Not only the tunable antennas, but also the antennas, whose shape is completely determined by the angular dimensions, conform to the principle of electrodynamics similarity. In this case changing of scale does not change the antenna, i.e., a radiator shape and dimensions in wavelengths are the same at different frequencies.

Antennas having the property of the automatic "cut-off" of currents excited in frequency-independent antennas a particular interest. This property means that the field at each frequency is radiated by a current along a small antenna section, which is named by the active area, and that the electric current outside the boundaries of this area is quickly attenuated. Here, coordinates and dimensions of radiated section are rigidly related with the wavelength. If the frequency is changed, the antenna section, radiating the field, shifts along the antenna. The electrical dimensions of the section (both longitudinal and cross dimensions) remain constant and ensure the invariability of the characteristics. Thus, the antenna has the constant input impedance and invariable directivity characteristics in an infinitely wide band.

If the antenna has finite dimensions, its frequency range is finite, but in this finite range the antenna has the properties of an infinite antenna. The maximal wavelength depends on the maximal cross dimension of the antenna (on its width), and the minimal wavelength depends mostly on the accuracy of the structure manufacturing near the excitation point.

LPDA (Fig. 11.7) is a set of elements (of wires), dimensions of which form a geometric progression with denominator $1/\tau$:

$$R_{n+1}/R_n = l_{n+1}/l_n = 1/\tau. \tag{11.24}$$

Here R_n is the distance from the vertex of the angle α to dipole n, l_n is the arm length of dipole n, α is the angle between the antenna axis and the line passing through the dipoles ends (see Fig. 11.7). Accordingly, the antenna electrical characteristics are repeated at frequencies forming the geometric progression with the same denominator. It means that directivity characteristics and input impedance of the antenna are periodic functions of a logarithm of frequency f, i.e., if the electrical characteristics are depicted as a function of $\ln f$, their values are repeated with period equal to $\ln \tau$. From here the antenna name was selected.

Weak variation of antennas characteristics within the period is a necessary condition of a weak frequency dependence of these characteristics. In order to meet this condition, this period must be small. But this is insufficient.

LPDA, shown in Fig. 11.7, consists of two structures located in the same plane. Each structure has the form of a straight wire with linear conductors attached to it at right angles alternately from the left and from the right. Their length increases with increasing distance from the excitation point in accordance with the law of geometric progression. Such an antenna is a simplified and modified version of the flat log-periodic structure shown in Fig. 11.8, which is a self-complementary structure, that is, consists of metal plates and slots coinciding with each other in shape and size. The input impedance of a flat infinite self-complementary structure is purely active, independent of frequency, and in the simplest case is equal to 60π Ohm (see Chapter 5). Designing a log-periodic antenna in the form of a self-complementary or similar structure provides a small change in its electrical characteristics during one interval of geometric progression.

Each of structures, forming *LPDA* (see Fig. 11.7), differs from the structure, which forms a flat log-periodic antenna (see Fig. 11.8). The metal sector 1' is replaced with the longitudinal wire 1, the metal strip 2' situated along the arc of a circumference is replaced with transverse wire 2, tangent to the arc, and the slot 3' is replaced with the interval 3 between the transverse wires. Such construction is dipole arms connected to wires of a straight long line. It is essentially simpler for implementation and, at the same time, its

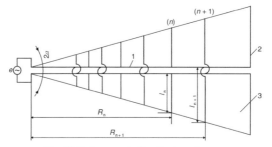

Fig. 11.7: Log-periodic dipole antenna.
1—longitudinal wire, 2—transverse wire, 3—interval between the transverse wires.

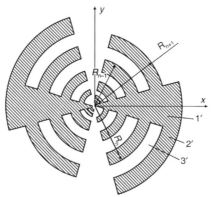

Fig. 11.8: Flat self-complementary log-periodic antenna.
1'—metal sector, 2'—metal strip, 3'—slot strip.

electrical characteristics are close to the electrical characteristics of original construction.

Rotation of one metal structure (of one antenna arm) around the *y*-axis (see Fig. 11.8) through angle π and placing both structures in one plane allow providing unidirectional radiation. The unidirectional log-periodic antenna shown in Fig. 11.7 may be interpreted as a linear array of symmetrical radiators. These radiators have monotonically changing lengths and are excited by a two-wire long line. A generator is connected to the line from the side of shorter radiators.

A reasonable implementation of an antenna design, which is requires no special balun, is shown in Fig. 11.9. The cable is placed inside one of two tubes forming a two-wire distribution line. The cable sheath and the tube of distribution line form a single unit, and an inner conductor of the cable is connected to the second tube at the antenna vertex. This design provides a shortcut of the distribution line. It is implemented at the distance $\lambda_{max}/8$ from the base of a first dipole. Here λ_{max} is the maximal wavelength.

Unfortunately, the opinion that the log-periodic structure itself provides constant input impedance is widespread. In [29] it is said that "there are serious misunderstandings. It seems, that such ignorance can be attributed to the term "log-periodic antenna" for the self-complementary log-periodic antenna, without reference to the most important fact that it is a derivative of the self-complementary structure … In order to correct such misunderstandings, experimental tests have been done by taking a conically-bent modified antenna, which is arranged in the log-periodic manner as shown in Fig. 11.10a (the figures numbers are replaced—B.L.). As the most straightforward arrangement of non-self-complementary log-periodic structure, the antenna shown in Fig. 11.10b was constructed, where one wing of the two half-structures of the antenna is upside-down … The measured values of input resistance for these two antennas are compared in Fig. 11.11, and a significant difference is apparent between them, in spite of the fact that two wings of both antennas are identical.

The input resistance of the incorrectly arranged log-periodic structure, which is shown in Fig. 11.11 by the dotted curve and crosses, varies distinctly in a log-periodic manner for varying frequency, though the constant-resistance property is satisfactory for the self-complementary antenna (its input resistance is shown by the solid curve and circles). From the results described above, it can be concluded that the origin of the broad-band property of the 'log-periodic antenna' is not in its log-periodic shape, but rather in the aspect of the shape that is derived from the self-complementary structure".

Hardly is it necessary to add anything to these words.

Further we consider the active area of *LPDA* with the view of explaining the principle of its operation. The area consists of dipoles with the arm

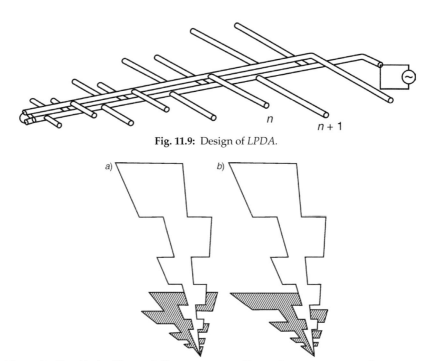

Fig. 11.9: Design of *LPDA*.

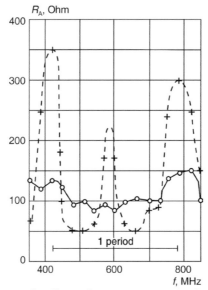

Fig. 11.10: Two kinds of log-periodic antenna: a—self-complementary antenna, b—antenna, which is built by anti-complementary method.

Fig. 11.11: Input resistances of a self-complementary antenna (ooo) and of an antenna, which is built by anti-complementary method (+++).

length close to $\lambda/4$. In their input impedance an active component is predominant, and the reactive component is small. In actual practice the number of dipoles, forming the active area, usually is equal to five. For the sake of simplification, we assume that there are only three dipoles, with the arm length of central dipole being $\lambda/4$.

As is seen from Fig. 11.7, the upper arms of dipoles connected alternately to one or another conductor of the distribution line. That is equivalent to crossing conductors of the long line on the sections between the dipoles. With allowance for this crossing the electrical current in a longer dipole outstrips in phase the current in the resonance radiator, and the current in a shorter dipole lags behind the current in the resonance radiator, i.e., the longer dipole acts as a reflector, and the shorter dipole acts as a director. As a result, the fields of individual radiators are summed in the direction toward the excitation point (to the side of shorter dipoles) and compensate each other in the opposite direction.

The waves in the distribution line, reflected from the dipoles of the active area, cancel each other to a large degree, since the reactive components of the input impedances of short and large dipoles are opposite in sign. This explains a high level of matching of the active area with the distribution line. In addition, the electrical length of the line from the feed point to the active area remains unchanged during the frequency change. Therefore, an impedance of active area transformed to the antenna input is the same at different frequencies as well.

The dipoles located outside the active area are excited weakly due to the great reactive impedance. The short dipoles at the beginning of structure practically fail to radiate, since the fields created by them summed almost in anti-phase because of crossing wires and the proximity of dipoles to each other (as compared with the wavelength). As a result, an electromagnetic wave along this section of line does not weaken, i.e., the distribution of currents and voltages at the line section between the excitation point and the active area is close to the mode of the traveling wave. The short dipoles act as capacitances shunting the distribution line and thereby decreasing slightly its wave impedance. The long dipoles situated behind the active area radiate weakly too, since, first, their input impedances are great and, second, the power of the wave at that section of line drops substantially as a result of attenuation in the active area.

The method of *LPDA* calculation [26] is based on antenna presentation in the form of a parallel connection of two multipoles (Fig. 11.12), one of which describes system of dipoles and is defined by matrix $[Z_A]$ of mutual impedances, and the other describes the distribution line and represents the matrix $[Y_l]$ of admittances. For each cross-section n of the structure, where

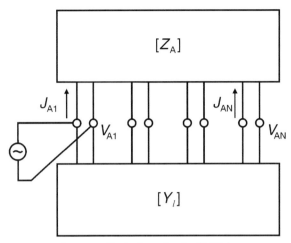

Fig. 11.12: Equivalent circuit of *LPDA*.

the dipole is connected in parallel with the distribution line the following equations are true:

$$J_{nA} Z_{nA} = J_{nl}/Y_{nl}, \quad J = J_{nA} + J_{nl}, \qquad (11.25)$$

i.e., $J_{nl} = J_{nA} Z_{nA} Y_{nL}$. Here J_{nA} is the current at the dipole input, Z_{nA} is the input impedance of the dipole (with allowance for coupling with neighboring dipoles), J_{nl} is the current of the distribution line, Y_{nl} is the admittance of the line in the cross-section n, and J is the extraneous current at given point. It should be noted that in calculating J_{nA} the mutual coupling with neighboring dipoles is accounted, and in calculating Y_{nl} it is considered that the distribution line is shorted at the terminals of neighboring dipoles (according to Kirchhoff's law other sources of emf are replaced by short circuit).

The first equation of (11.25) is written for the total voltage along closed circuit, the second equation is written for the total current in this point. From here,

$$J = (1 + Z_{nA} Y_{nl}) J_{nA}. \qquad (11.26)$$

Accordingly, a matrix equation for the column-vector $[J_A]$ of the dipole's input current is written in the form

$$[J] = ([E] + [Z_A][Y_l])[J_A], \qquad (11.27)$$

where $[E]$ is the identity matrix, $[J]$ is the column-vector of currents feeding the lines, which connects multipoles with each other. Since the extraneous current is only at the distribution line input, in the first cross-section (it is equal to J_0), then

$$[J] = \begin{vmatrix} J_0 \\ 0 \\ \cdots \\ 0 \end{vmatrix}.$$ Solving equation (11.27), we find the column-vector $[J_A]$, and

then matrix $[V_A] = [Z_A][J_A]$ of the voltages at the dipoles inputs. The first element of the matrix at the input of the shortest dipole is the voltage. If the exciting current J_0 is equal to 1, this first element is equal to the input impedance of the antenna.

In [26], the elements of the matrix $[Z_A]$ are calculated, in fact, by means of the induced emf method. Later on, to obtain more exact results, the matrix elements were calculated by means of the integral equation's solution with the help of the moment method [59]. The difference between the approximate and the exact method is particularly noticeable, if the *LPDA* consists of thin radiators or has a wide angle at the antenna's vertex. The energy in such antenna propagates along the distribution line beyond the boundaries of the active area and excites the long dipoles.

When designing *LPDA* it is important to choose the geometric dimensions so that the electrical characteristics changed weakly in a range from f to τf. The magnitude τ and all antenna characteristics depend essentially on the parameter σ, which is equal to the distance between the half-wave dipole and the neighbor shorter dipole (in wavelengths):

$$\sigma = 0.25(1 - \tau) \cot \alpha. \qquad (11.28)$$

In fact, it is dependence on the angle α with allowance for τ. As it is shown in [26, 60], the characteristics change weakly, if $\tau > 0.8$ and $0.05 \leq \sigma \leq 0.22$. Under these conditions, the currents of the dipoles located near the resonant (half-wave) radiator reach a maximum and the wave along the distribution line is so attenuated in the active area that the follow dipoles practically do not radiate.

In [61] on the basis of generalization of data available in the literature the optimum relationship of the above mentioned basic parameters is defined in the form:

$$\sigma/\tau = 0.191. \qquad (11.29)$$

This ratio does not depend on the values of α, l_n/a_n and Z_0. Here a_n is the radius of dipole n, $Z_0 = 60 \, ch^{-1}[(D^2 - 2a^2)/(2a^2)]$ is the wave impedance of the distribution line, a is the radius of the distribution line's wires, and D is the distance between axes of these wires. Substituting (11.29) into

(11.30), authors of [61] obtain the simple expressions connecting the optimal parameters τ and σ with the antenna dimensions:

$$\tau = 1/(1 + 0.765 \tan \alpha) = L/[L + 0.765(l_1 - l_N)], \ \sigma = 1/(4 \tan \alpha + 5.23). \quad (11.30)$$

The value L in these expressions is the distance between the first and the last (N) dipole.

The antenna with $\sigma/\tau = 0.191$ has a narrow directional pattern and high front-to-back ratio. Figures 11.13 and 11.14 corroborate these statements. They show the given in [61] calculated beam width for $LPDA$ with $Z_0 = 100$ Ohm and $l_n/a_n = 177$ in the planes E and H and also front-to-rear ratio depending on the parameters τ and σ. SWR of the same antenna with the optimal parameters τ and σ depending on the values l_n/a_n and Z_0 is presented in Figs. 11.15 and 11.16. Magnitude of SWR in a properly designed $LPDA$ is typically smaller than 1.5.

Under antenna development it is necessary to take into account that the arms of each dipole are connected to different conductors of the distribution line, and so they are not coaxial. To decrease the influence of misalignment on the antenna pattern, one must reduce the distance between the conductors' axes: it should not exceed $0.02 \ \lambda_{min}$. Here λ_{min} is the minimum wavelength.

Log-periodic antennas have rather large overall dimensions. In order to decrease transverse dimensions, it is expedient to shorten the longest dipoles using loads of different kind or structures with the slowing-down, i.e., the same manners, which are used for reducing the monopole's and dipole's length. Different variants of shortened monopoles are presented in Fig. 11.17. Among them, inverted L and T-radiators (a, b) and antennas with concentrated inductive loads (c) are. Slowing-down is employed in a helical (d) and meandering (e) antennas and in monopoles of fractal shape of Koch (f). It should be noted that the slowing factor is always less than the increase of the wire length.

Slowing-down allows shortening the monopole, i.e., to reduce the length of the monopole in m times for the given frequency of the first resonance or to decrease the resonance frequency in m times for the given length of the monopole. But the radiation resistance at the resonance frequency in consequence of the length reduction decreases in m^2 times, and the antenna wave impedance is increased in m times. And both impair matching with the cable of each element of $LPDA$ and the antenna on the whole.

Figure 11.18 demonstrates the results of a rigorous calculation of SWR and gain for log-periodic antennas with linear and helical dipoles. Parameters of the antenna are following: $N = 15$, $\tau = 0.92$, $\alpha = 10°$, $l_n/a_n = 100$, $l_n/\rho_n = 20$ (here ρ_n is the radius of helical dipole n), $Z_0 = 150$ Ohm. The helical dipole arm consists of five turns; the wire length is twice as large than the straight dipole's length. The value of SWR is calculated in a

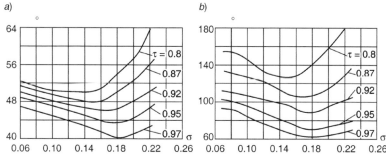

Fig. 11.13: Dependence of half-power beam width of *LPDA* with $Z_0 = 100$ Ohm и $l_n/a_n = 177$ in the planes E (a) and H (b) on the parameters τ and σ.

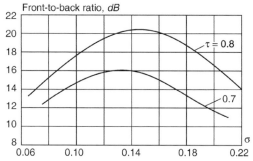

Fig. 11.14: Dependence of front-to-back ratio on the parameters τ and σ.

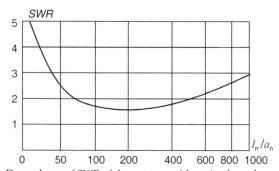

Fig. 11.15: Dependence of SWR of the antenna with optimal τ and σ on value l_n/a_n.

cable with wave impedance 100 Ohm. The relative length l_N/λ of the largest dipole's arm is used as the argument.

As one can be seen from Fig. 11.18, the level of $TWR \geq 0.7$ for the antenna with the helical dipoles is maintained in the range $0.163 \leq l_N/\lambda \leq 0.425$. The dotted curves in the figure correspond to the log-periodic antenna with straight dipoles. The figure shows that, if both antennas have the same dimensions, *LPDA* with the helical dipoles and a double wire length has an operation range, expanded by half in the direction of low

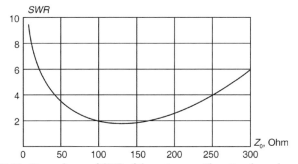

Fig. 11.16: Dependence of SWR of the antenna with optimal τ and σ on Z_0.

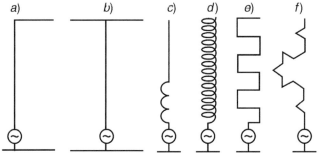

Fig. 11.17: Variants of shortened monopoles: a—inverted-L antenna, b—T-antenna, c—antenna with concentrated load, d—helical antenna, e—meandering antenna, f—antenna of Koch fractal shape.

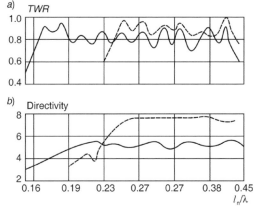

Fig. 11.18: Characteristics of the log-periodic antennas with helical dipoles (solid curve) and straight dipoles (dotted curve): a—traveling-wave ratio, b—directivity.

frequencies in comparison with the range of ordinary antenna. The useful effect is accompanied by decreasing match level and some deterioration of directivity, caused by a higher Q of helical dipoles.

If to increase in the considered example the parameter τ up to 0.95 and the number of dipoles up to 24, we shall obtain the antenna, the characteristics of which are almost the same as the characteristics of an antenna with straight dipoles and the transverse dimensions reduced by half. Thus, it is theoretically possible to reduce its transverse dimensions at the cost of increasing number of dipoles and at the same time to maintain characteristics of the log-periodic antenna. But practically acceptable designs are obtained, if the transverse dimensions are reduced no more than two or three times.

Attempts to decrease longitudinal dimensions of an antenna by using slowing-down in the distribution line or at the expense of additional dipoles connection, failed, since violation of relationships corresponding to geometric progression and increase of number of dipoles causes, as a rule, sharp deterioration of electrical characteristics and gives insignificant decrease of overall dimensions.

The variant of log-periodic antenna, which operates in two adjacent frequency bands and allows making the antenna shorter than antenna designed for operation in the total range, is described in [61]. Basically, the authors' proposal reduces to the use of linear-helical dipoles, i.e., radiators, each of which consists of straight and helical dipoles arranged coaxially and having the common feed point (Fig. 11.19).

The dipoles length is the same, but the helical wire length is twice as much as the straight rod's length. Linear-helical dipole in contrast to straight and helical dipole has two serial resonances, and the ratio of the resonant frequencies for the same dipole's length is equal to the slowing factor of the helical dipole.

As is well known, resonant dipole and its nearest neighbors create an active area, passing through which the electromagnetic wave, whose frequency is close to the resonant frequency, actively radiates energy. *LPDA* with linear-helical dipoles has two active areas, and they provide a signal radiation in two bands of the frequency range. The experimental check of log-periodic antenna with linear-helical dipoles, described in [61], confirms that this proposal is promising. The antenna is designed for operation in the frequency range from 250 to 1250 MHz. The length of mock-up is equal to 0.44 m; the dipole maximum length is 0.42 m. The test results are given in Fig. 11.20–11.23.

From Fig. 11.20 it is seen that the *TWR* in the cable with wave impedance 75 Ohm is greater than 0.3 in the ranges 252–610 and 645–1250 MHz, at the frequencies 613 and 625 MHz its value decreases to 0.17 and 0.18, respectively. The front-to-back ratio is greater 8 db (see Fig. 11.21). The half-power beam width (both in the plane E and in the plane H) in the lower part of range is wider than in the upper (see Fig. 11.22). Accordingly, here

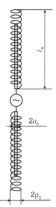

Fig. 11.19: The linear-helical dipole.

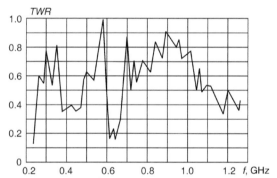

Fig. 11.20: *TWR* of antenna with linear-helical dipoles.

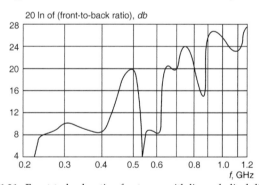

Fig. 11.21: Front-to-back ratio of antenna with linear-helical dipoles.

the antenna directivity is smaller. Directional radiation exists from 260 to 1250 MHz (see Fig. 11.23). Only at 550 MHz this ratio falls sharply to 2 db.

If electrical characteristics of log-periodic antenna with straight dipoles and with linear-helical dipoles are similar, then the length of the antenna with straight dipoles is greater in 1.8 times. If only helical dipoles are used,

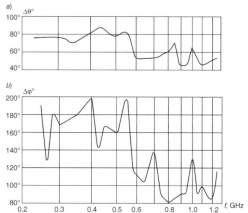

Fig. 11.22: Pattern of antenna with linear-helical dipoles in the plane E (a) and in the plane H (b).

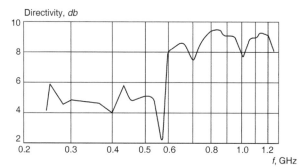

Fig. 11.23: Directivity of antenna with linear-helical dipoles.

the length of antenna is greater than length of antenna with linear-helical dipoles in 1.3 times. In addition, TWR of the proposed antenna in the upper part of the range is smaller on the average by 4 db. Decrease of the antenna dimensions is obtained at the cost of *TWR* and directivity reduction in the narrow band in the middle of the operation range. This reduction is caused by the transfer of the active region from the helical elements of *LPDA* to straight elements.

The length of log-periodic antenna can be reduced by increasing the angle α between the antenna axis and the line passing through the dipoles ends. This option seems the most simple and natural. But, as it is seen from (11.28), increase of α, if τ is constant, leads to decrease of the distance between the dipoles and to the growth of their mutual influence, and as a result to decrease of directivity and active component of input impedance and to the deterioration of the frequency-independent characteristics.

One can increase the angle α by another manner. *LPDA* consists (see Fig. 11.7) of two asymmetric structures located in the same plane and excited in opposite phases. If these structures to locate at an angle $\psi > \alpha$ to each other, as it is shown in Fig. 11.24, the resulting three-dimensional structure will incorporate two distant from each other planar structures. The monopoles are connected alternately from left and from right to the conductor of the distribution line. The distance between the monopoles, situated on the one side of the conductor, is almost twice as large as in a planar *LPDA*. This reduces their mutual influence and allows to increase the angle *a*. However, this antenna occupies a great volume, and that makes difficult its installation and changes its characteristics. This change, for example, an increase of input resistance, creates additional problems for antenna's utilization.

Asymmetrical coaxial log-periodic antenna, described in [61], has not these disadvantages. Two-wire distribution line in this antenna is replaced by a coaxial line, and dipoles are replaced by monopoles. Antenna as an assembly is shown in Fig. 11.25. The antenna consists of two structures, circuits of which are given in Fig. 11.26. The first of them (Fig. 11.26a) is a straight conductor. The wires' sections of required length located in one plane connected to defined points of this conductor at the right angle alternately from left and from right. This conductor is the central wire of the coaxial distribution line and the wires' segments are monopoles, which are excited by means of this conductor.

The second structure (Fig. 11.26b) is designed as a long cylindrical tube with embedded in it short tubes, which are opened inwards and out. The long tube is the outer shell of coaxial distribution line, the short tubes are the outer coaxial shells, which located around monopoles connected to the inner conductor of the distribution line. As a result, monopoles are the radiators with a feed point displaced from the base. As one can be seen from Fig. 11.26, the first structure is inserted into the second one, so that their axes coincide.

In accordance with the usual practice of designing log-periodic antenna, its dimensions must correspond to the geometric progression with ratio $1/\tau$:

$$R_{n+1}/R_n = l_{n+1}/l_n = h_{n+1}/h_n = 1/\tau, \tag{11.31}$$

where h_n is the distance from the axis of the distribution line to a feed point of radiator *n*. Other values are defined previously. In addition, the ratio of the coaxial section diameter to the central conductor diameter of each monopole should provide in this section coincidence of wave impedance with the radiation resistance on the frequency of first series resonance.

From the above it follows that the two-wire distribution line is replaced in the proposed antenna by a coaxial cable, and dipoles are replaced by monopoles connected to the inner conductor of this cable. The outer shell of the cable is used as a ground. This shell in turn serves as a ground for

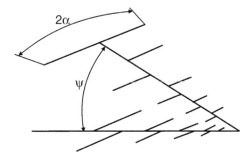

Fig. 11.24: Volumetric antenna of two structures.

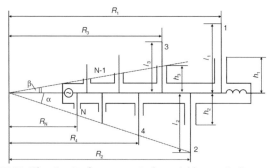

Fig. 11.25: The circuit of asymmetrical coaxial log-periodic antenna.

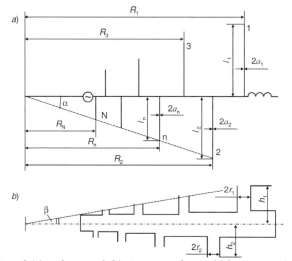

Fig. 11.26: Internal (a) and external (b) structures, from which asymmetrical coaxial log-periodic antenna consists.

monopoles excited in anti-phase, that substantially distinguishes this ground from large metal sheet. This means that the proposed structure realizes an asymmetrical version of the usual log-periodic antenna (symmetrical version of such antenna is implemented as the antenna *LPDA*). In such antenna one can significantly increase the angle α and decrease the length without fear of directivity decrease and deterioration frequency-independent characteristics. Since the radiating elements of the antenna are the monopoles, then by analogy with the *LPDA*, where the dipoles play a similar role, it is expedient to name this antenna by *LPMA*.

The principle of a symmetrical antenna's operation was reviewed earlier by means of the analysis of processes in its active area. The processes in the active area of an asymmetrical antenna are practically do not differ, since the waves in a coaxial distribution line are similar to the waves in the two-wire line and are depended on the monopole mutual influence, which is similar to the influence of dipoles in a symmetrical structure. In the surrounding space the equally excited dipoles and monopoles produce the same fields.

Mock-up of an asymmetric log-periodic antenna designed in order to operate in the range of 200–800 MHz, has been manufactured and tested. Antenna characteristics were measured by authors for the two variants of its mounting on the metal mast (Fig. 11.27): a—the cantilever variant, when the radiators are arranged vertically, b—installation on the mast top, where the radiators are mounted horizontally, and the gravity center coincides with the mast axis. Distribution line was formed in the shape of a truncated pyramid with a square cross-section and the inner conductor made in the shape of a horizontal plate of variable width.

Experimental check confirmed that *LPMA* regardless of the installation variant has frequency independent electrical characteristics. The cantilever variant gave the following results. The average magnitude of half-power beam width in the operation range 200–800 MHz is equal to 70° in the plane *E* (vertical) and 124° in the plane *H* (horizontal). Typical patterns in both planes are shown in Fig. 11.28. Back-to-front ratio does not exceed 0.15. The directivity value is 6.8 db.

The experiment shows that the antenna does not lose the directional characteristics up to frequency 2.75 GHz. In Fig. 11.29 *TWR* of antenna is given in the cable with the wave impedance 50 Ohm in the range from 0.8 to 2.8 GHz. One can be seen that in the range from 1.5 to 2.4 GHz *TWR* is equal to 0.3–0.7. The beam width in the plane *E* in this range is 35–40° (i.e., half as much than that in the main operation range), and in the plane *H* it is equal to 115–140°. Accordingly, in additional range directivity is higher by a half than in the main range.

The authors' point of view on the causes of additional operating range's emergence is absent in [61]. From our point of view, the reason is obvious enough, if to take into account the calculations and measurements results

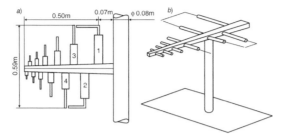

Fig. 11.27: Asymmetrical antenna on the metal mast: a—cantilever variant, b—installation on the mast top.

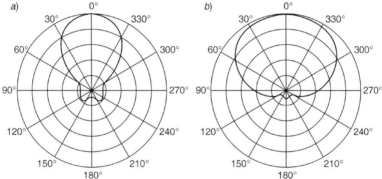

Fig. 11.28: Typical directional pattern of asymmetrical antenna in the plane *E* (a) and in the plane *H* (b).

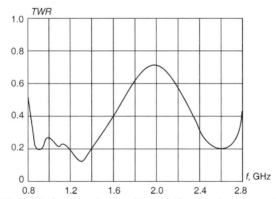

Fig. 11.29: *TWR* of antenna in the cable with the wave impedance 50 Ohm.

for *LPDA* with linear-helical dipoles presented by the authors. Here each radiator consists of two connected in parallel elements with different resonant frequencies. Actually, in this *LPMA* along the distribution line two monopoles' structures are set. The dipoles' dimensions are defined by two different angles—α and β (see Fig. 11.26)—between the axis of the antenna and the line passing through the ends of the radiators of both structures. Each structure provides the required electrical characteristics within its operating range. The sharp deterioration of characteristics occurs at the boundary of the ranges.

The similar result during measurements of *LPMA* is caused by the fact that each radiator of the antenna consists of two connected in parallel elements: the monopole and the short tube, i.e., the section of coaxial cable, surrounding the monopole. Structure dimensions are chosen so that the distance from the central conductor axis to the end of the tube is equal to half of the monopole length. But that does not mean that the length of a single element is half of the other element's length because the length of one element is equal to the short tube's length (it is necessary to subtract the radius of the distribution tube). Therefore, the average frequency of the additional range (1.95 GHz) is greater approximately four times (not two times) than average frequency of the main range (0.5 GHz).

When the mock-up of *LPMA* is placed on the mast top, *TWR* in the cable with a wave impedance 75 Ohm does not fall below 0.3 in the range from 115 to 800 MHz (Fig. 11.30), i.e., the lower frequency decreased by a factor 1.6. Since for the sake of decreasing the transverse dimensions of the antenna the first monopole was bent at right angle at the height of the third monopole and the second monopole was bent at the height of the fourth monopole, the third and the fourth monopole have the largest length. At a frequency 115 MHz, the length of the third and fourth monopole is chosen equal to 0.227 λ.

Typical directional pattern of antenna placed on the mast top, in the *E* plane at frequencies above 300 MHz is similar to the directional pattern shown in Fig. 11.22. When the frequency decreases, the back lobe grows. At frequencies below 160 MHz, the directional pattern with two lobes turns and becomes perpendicular to the distribution line axis. This rotation is caused by the fact that at low frequencies the outer tube of the distribution line becomes the main radiator.

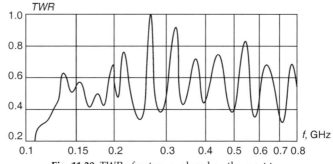

Fig. 11.30: TWR of antenna, placed on the mast top.

12

Different Issues

12.1 Antenna with arbitrary single load

The problem under study is connected with an unexpected change of an antenna structure and with an analysis of the consequences of this change (one can name such change by emergency). In general case, this emergency change can be considered as the appearance of an arbitrary load at an arbitrary point of the antenna. Obviously, the result of this event depends not only on a load magnitude, but also on the antenna structure.

The problem of loads effect on antennas characteristics has a general nature, but it manifests by different variants. Loads can be concentrated and distributed, i.e., their dimensions may be small and comparable with a wavelength. Among them there are for example nonlinear loads [62]. They include non-linear contacts and pieces of wires from superconducting metal, as well as contacts between wires, damaged for various external reasons. Finally, radio elements, specially included in an antenna circuit, both active (resistors) and reactive (capacitors, inductors), may also serve as loads. In Fig. 12.1 circuits of two straight symmetrical radiators are shown: of a metal radiator with sinusoidal current distribution (a) and of a radiator with in-phase current (b), created by capacitive loads. Arbitrary complex impedances are included in each antenna arm.

An integral equation for a current in an antenna with capacitive loads may be obtained from an integral equation for a current $J_A(z)$ in a metal radiator. An inclusion of a concentrated complex impedance Z at a certain point $z = h$ is equivalent to an inclusion at this point of an additional emf $e_1 = -J_A(h)Z$. An extraneous field, which corresponds to additional emf, is equal to a product of the current at this point into the mentioned impedance:

$$K_1(z) = -J_A(h)Z\delta(z - h). \tag{12.1}$$

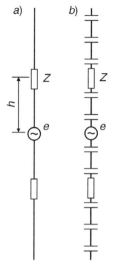

Fig. 12.1: Symmetrical radiators with sinusoidal (a) and in-phase (b) currents and with arbitrary complex impedances.

The boundary condition for an electric field $E_z(J)$ on a radiator surface with n loads is written as

$$E_z(J_A)\big|_{\rho=a,-L\leq z\leq L}+K(z)-\sum_{n=1}^{N}J_A(h_n)Z_n\delta(z-h_n)=0. \qquad (12.2)$$

Here a cylindrical coordinate system is used, a is an antenna radius, and L is its arm length. If only one load was created in the radiator, then in accordance with (12.2) the Leontovich's integral equation takes a form:

$$\frac{d^2J_A(z)}{dz^2}+k^2J_A(z)=-4\pi j\omega\varepsilon_0\chi\left[K(z)+W(J_A,z)J_A(z)-J_A(h)Z\delta(z-h)\right], \qquad (12.3)$$

where k is the wave propagation constant along the antenna, $\chi=1/[2\ln(2L/a)]$ is a small parameter, $W(J_A,z)$ is the functional. The right side of equation (12.3) contains three summands in square brackets: first summand allows to take into account the exciting emf e, second one—the radiation, third one—the load presence. The equation solution, as usual, is sought in the form of a series in powers of the parameter χ:

$$J_A(z)=\chi J_1(z)+\chi^2 J_2(z)+\dots, \qquad (12.4)$$

that permits to arrive at a system of equations for the members of series. If Z is a magnitude of order $1/\chi$, i.e., it is comparable with the antenna wave impedance, then the equation for the current $J(z)=\chi J_1(z)$ of the first order of smallness has the form

$$\frac{d^2 J(z)}{dz^2} + k^2 J(z) = -4\pi j\omega\varepsilon_0\chi\left[K(z) - J(h)Z\delta(z-h)\right], J(\pm L) = 0. \quad (12.5)$$

Substituting the value $z = h$ in the solution of equation (12.5), we find (see [35]) for an antenna excited at the center,

$$J(z) = j\frac{\chi e}{15\cos kL}\sin k(L - |z|) + \frac{\chi^2 e}{900\sin^2 2kL}\frac{ZZ_1}{Z + Z_1}\sin kL \sin k(L + \gamma h)\sin k(L - |h|)\sin k(L - \gamma z),$$
$$(12.6)$$

where $Z_1 = -j\dfrac{30\sin 2kL}{\chi \sin k(L+h)\sin(L-h)}, \gamma = \begin{cases} 1, z \ge h \\ -1, z \le h \end{cases}$. As this was to be expected, the current along the antenna with one concentrated load contains two sinusoidal terms: one of them is created by the generator, and the other is caused by the load presence.

An example of affecting a concentrated load on a current distribution in a metal radiator, i.e., in an antenna with a sinusoidal current distribution, confirms that an inclusion of the concentrated complex impedance Z at the point $z = h$ is equivalent to placing in it an additional concentrated emf $e_1 = -J_A(h)Z$. The distribution of current, generated by this emf, depends on the antenna structure, and a magnitude of generated current at the point of placing load depends on the input impedance in this point. In a straight metal radiator this input impedance in the first approximation is a reactance, equal to

$$jX_M = -jWctgk(L - h) - jWctgk(L + h), \quad (12.7)$$

where $W = 30/\chi = 60\ln(2L/a)$ is the wave impedance of each antenna arm; $L - h$ and $L + h$ are the arms' lengths.

In radiators with in-phase currents created by connecting capacitive loads into the antenna wire, the input impedance changes substantially. If the capacitances of loads decrease proportionally to the distance to the free end of the antenna then a current diminishes linearly, and the input impedance is

$$jX_{IF} = -j\frac{W_{IF}}{k(L-h)} - j\frac{W_{IF}}{k(L+h)L}. \quad (12.8)$$

where $W_{IF} = \gamma_n W/k$, γ_N is the wave propagation constant at the nearest antenna section, and W is the wave impedance of the antenna with the same dimensions, but without loads.

It is useful to compare the fields created by symmetrical radiators with sinusoidal and in-phase currents. The field of a symmetrical vertical radiator is

$$E_\theta = j30k\sin\theta\frac{e^{-jkR}}{R}\int_{-L}^{L}J(z)e^{jkz\cos\theta}dz,\tag{12.9}$$

where R is the distance from the antenna center (from coordinates origin) to the observation point, $kz\cos\theta$ is the path difference between the antenna center and the point z to an observation point. The current magnitude is symmetrical about the origin. Therefore

$$E_\theta = 2A\sin\theta\int_{0}^{L}J(z)\cos(kz\cos\theta)dz,\tag{12.10}$$

where $A = j30ke^{-jkR}/R$. In a case of a sinusoidal distribution $J(z) = J(0)\sin k (L - z)$. Using a known equality $2\sin\alpha\cos\beta = \sin(\alpha - \beta) + \sin(\alpha + \beta)$, we obtain for the total field of the radiator

$$E_\theta = A\sin\theta\left\{\frac{\cos k[L-z(1+\cos\theta)]}{k(1+\cos\theta)}+\frac{\cos k[L-z(1-\cos\theta)]}{k(1-\cos\theta)}\right\}\Big|_0^L=\frac{2A}{k\sin\theta}[\cos(kL\cos\theta)-\cos kL]\cdot\tag{12.11}$$

The appearance of identical concentrated loads at points h and $-h$, as already mentioned, is equivalent to inclusion of two generators at these points with emf $e_1 = e_2 = -J_A(h)Z$, where $J_A(h) = J(0)\sin k(L - h)$ is a current magnitude at the point of the load inclusion. The output current of each generator is equal to $J_2(h) = -J_A(h)Z_1/(jX_M)$, i.e.,

$$J_{1,2}(h) = e_{1,2}/(jX_M) = -J(0)\sin k(L - h)Z/(jX_M).\tag{12.12}$$

Each generator creates currents in wires, located between it and the antenna ends. Their lengths are equal to $L - h$ and $L + h$. The current in the short wire is equal to $J_1(z) = J_1(h)\sin k(L - z)/\sin k(L - h)$, in the long one $J_2(z) = J_2(h)\sin k(L - z)/\sin k(L + h)$. The field of additional radiators is in accordance with (12.11)

$$E_\theta = \frac{2B}{k\sin\theta\sin k(L-h)}[\cos(kh\cos\theta)-\cos kh+\cos(k\langle L+h\rangle\cos\theta)-\cos k(L+h)],\tag{12.13}$$

where $B = -AZ/[jX_AJ(0)]$.

In an antenna with an in-phase current distribution $J(z) = J(0)(1 - z/L)$, i.e., the total field of such a radiator is

$$E_\theta = A\sin\theta\left\{\frac{\sin(kz\cos\theta)}{k\cos\theta}-\frac{1}{L}\left[\frac{\cos(kz\cos\theta)}{k^2\cos^2\theta}+\frac{z\sin(kz\cos\theta)}{k\cos\theta}\right]\right\}\Big|_0^L=\frac{2A\sin\theta}{k^2L\cos^2\theta}[1-\cos(kL\cos\theta)]\cdot\tag{12.14}$$

In the presence of loads, the field of additional radiators

$$E_\theta = \frac{2B\sin\theta}{k\cos^2\theta}\left\{\frac{1}{kh}\left[1-\cos(kh\cos\theta)\right]+\frac{1}{k(L+h)}\left[1-\cos(k\langle L+h\rangle\cos\theta)\right]\right\}. \quad (12.15)$$

The described method is used for calculating electrical characteristics of antennas with sinusoidal and in-phase current distribution, as well as characteristics of the same antennas in a case of connecting concentrated loads that substantially change the antennas currents. The calculation procedure is described in Chapter 4. Specific calculations are performed using Excel.

In Fig. 12.2 directivity D and pattern factor PF of radiators with in-phase (1) and sinusoidal (2) currents, depending on the arm length of the antenna are presented for the case, when arbitrary single loads are absent. As can be seen, the working range of radiators with in-phase currents is much wider.

The results of calculating fields of each antenna without arbitrary loads and in a presence of these single loads (resistors with $R = 1000$ Ohm) are given in Table 12.1. They allow to compare the effect of loads on the characteristics of different radiators, depending on their magnitudes. The Table 12.1 uses the following notation: E_θ is a total field of antenna without the arbitrary load, ΔE_θ is a total field of antenna with this load, $\Delta E_\theta/E_\theta$ is the ratio of two these fields, whose big value indicates a substantial change of characteristics, an unstable nature and a significant effect of arbitrary load on the magnitudes and distributions of the current. The Table 12.1 shows the length kL of the antenna arm and the distance kh from this load to the antenna center. In the first part of Table 12.1 these values are resonant, and this leads to a very sharp change of fields, especially fields of the antenna with sinusoidal current distribution. In the second part of Table 12.1 the values kh are changed, in the third part similar changes are made in antenna sizes. As a result, the change of the antennas fields became weaker, but the overall result does not change.

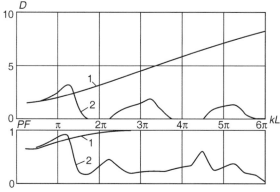

Fig. 12.2: Directivity D and pattern factor PF of radiators with in-phase (1) and sinusoidal (2) currents in the absence of arbitrary load.

Table 12.1: Fields of antennas.

kL	In-phase currents			Sinusoidal currents		
	E_θ	ΔE_θ	$\Delta E_\theta/E_\theta$	E_θ	ΔE_θ	$\Delta E_\theta/E_\theta$
$h = 0.5L$						
$0.25\,\pi$	0.62	0.13	2.1	0.056	1.18	21.1
$0.5\,\pi$	0.94	1.7	0.8	0.61	$3.1 \cdot 10^{11}$	$5 \cdot 10^{11}$
$0.75\,\pi$	4.33	6.2	1.4	1.55	9.8	6.5
π	12.2	14.4	1.2	1.66	$1.64 \cdot 10^{11}$	10^{11}
$1.25\,\pi$	26.4	27.4	1.04	0.89	3.6	4.0
$1.5\,\pi$	48.3	47	0.97	0.88	$1.44 \cdot 10^{10}$	$1.64 \cdot 10^{10}$
$1.75\,\pi$	79.3	74.9	0.94	1.92	17.3	9.0
$2\,\pi$	121	112	0.93	2.16	$6 \cdot 10^{-9}$	$3 \cdot 10^{-9}$
$h = 0.3L$						
$0.25\,\pi$	0.62	0.2	3.2	0.056	1.55	27.7
$0.5\,\pi$	0.94	2.71	2.88	0.61	$8.44 \cdot 10^{11}$	$13.8 \cdot 10^{11}$
$0.75\,\pi$	4.33	10.6	2.45	1.55	0.86	0.55
π	12.2	25.1	2.06	1.66	$2.1 \cdot 10^{11}$	$1.3 \cdot 10^{11}$
$1.25\,\pi$	26.4	47.3	1.79	0.89	7.0	7.86
$1.5\,\pi$	48.3	79.2	1,64	0.88	$5.6 \cdot 10^{7}$	$6.4 \cdot 10^{7}$
$1.75\,\pi$	79.3	123.4	1.56	1.92	8.4	4.38
$2\,\pi$	121	182	1.5	2.16	$3.9 \cdot 10^{10}$	$1.8 \cdot 10^{10}$
$h = 0.5L$						
$0.15\,\pi$	0.0082	0.0184	2.2	0.0079	0.051	6.46
$0.3\,\pi$	0.128	0.27	2.13	0.11	3.9	35.5
$0.45\,\pi$	0.62	1.18	1.9	0.44	176	400
$0.6\,\pi$	1.88	3.1	1.68	1.0	15.3	15.3
$0.75\,\pi$	4.33	6.2	1.43	1.55	9.8	6.3
$0.9\,\pi$	8.4	10.7	1.27	1.77	110.6	62.5
$1.05\,\pi$	14.5	16.6	1.14	1.53	319	208.5
$1.2\,\pi$	23	24.3	1.06	1.04	10.4	10

Calculation shows that the use of radiators with in-phase linear current distribution makes it possible to significantly decrease the arbitrary loads effect on the antenna characteristics. This result naturally complements the results of researching antennas with in-phase currents. The analysis confirms a wider range of such antennas, an increase of directivity and a pattern factor, a scope for reduction of antennas dimensions, as well as the advantages of new methods for comparing antenna characteristics in accordance with current distribution (see Chapter 4).

12.2 Transparent antennas

A transparent antenna may be stand an example of using self-complementary structures consisting of several metal and slot radiators. In this antenna, the described structures allow to provide the required level of the antenna matching with the signal source.

Let's start with the necessary information about transparent antennas. The creation of such antennas became possible thanks to a working out of thin transparent and conductive films. Such antennas have unconditional advantages. First, they can be made invisible. Secondly, they can be used in the capacity of screens for projecting different images—both still (photos) and moving images (for example, *TV*). This option of additional use is especially important for small devices where antennas are installed, that is, for operation at high radio frequencies.

Thin films of ITO (Indium-Tin-Oxide), placed on high-quality glass substrates, are electrically conductive and optically transparent at ultrahigh and superhigh frequencies. They have a high uniformity of surface impedance. This allows to use them as flat antennas for mobile communications and other applications. The optical transparency of the *ITO* film for different resistivity is presented, for example, in Fig. 12.3, given in [63]. As can be seen from this figure, the transmission coefficient increases with increasing film resistivity and provides a sufficiently high transparency (about 95%) if the film resistivity is greater than 5 Ohm/square.

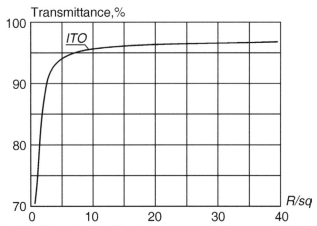

Fig. 12.3: Example of the film transparency at the frequency 0.545 GHz.

To better understand the material constraints imposed by the low conductivity of the *ITO* film, we will consider the surface resistivity of the film as a function of its thickness d. This quantity is denoted as R_{sq1}. According to Leontovich's boundary condition, if the thickness d of the metal film is greater than the penetration depth s, the film resistivity is equal to

$$R_{sq1} = R_{sq} = 1/(\sigma\delta) \text{ Ohm},\qquad(12.16)$$

where σ is its specific conductivity with respect to direct current (in S/m), and the penetration depth δ is given by the formula

$$\delta = 1/\sqrt{\pi f \mu \sigma},\qquad(12.17)$$

where f is frequency (in Hz), $\mu = \mu_0 = 4\pi \cdot 10^{-7}$ F/m is the absolute permeability. If the thickness d of the metal film is much smaller than the penetration depth δ, the film's sheet resistivity is equal to

$$R_{sq1} = R_{sq}\delta/d = 1/(d\sigma).\qquad(12.18)$$

The specific resistivity of *ITO* films is substantially greater than the specific resistivity of printed cards and metal antennas, where copper and aluminum are used. For example, the specific resistivity R_{sq1} of the transparent film CEC005P is equal to 4.5 Ohm/square. The specific conductivities of copper and aluminum are respectively $5.8 \cdot 10^7$ and $3.5 \cdot 10^7$ S/m, and hence in accordance with (12.16) and (12.17) the specific resistivity of a copper plate, whose thickness is greater than the penetration depth, at frequencies 1 and 5 GHz is equal to $6.9 \cdot 10^{-3}$ and $18.4 \cdot 10^{-3}$ respectively (the specific resistivity of an aluminum plate with analogous thickness at these frequencies is equal to $4.2 \cdot 10^{-3}$ and $11.1 \cdot 10^{-3}$ respectively). That means, the resistivity of *ITO* transparent film is greater by several orders than the resistivity of copper and aluminum, i.e., the conductive films are different from materials (copper, aluminum) commonly used in antennas by decreasing conductivity that changes properties of radiators.

For comparative analysis of antennas made of materials with high and low conductivity, it is necessary to apply methods for solving the corresponding boundary value problems of electrodynamics. These methods are divided into direct numerical and approximate analytical. The conclusion of specialists on the results of these methods application in prolonged researches is clearly formulated in [64]: "The undoubted advantage of analytical methods is that they are physically clearer in comparison with numerical methods. Analytical methods allow to determine the effect of device parameters on its individual characteristics".

In recent years transparent antennas have been the subject of many works [65–67]. However, these works were devoted only to definition and improvement of materials properties. Physical processes in transparent antennas, their electrical characteristics and their difference from characteristics of metal antennas with a high conductivity as a rule were not considered.

For understanding physical processes in an antenna knowledge of current distribution law along its axis has great importance. This knowledge allows defining all main characteristics of antennas and serves as the basis for the analysis of any antenna. This postulate remains valid in spite of elaboration of such calculation programs as program CST, since firstly these programs basically allow calculation of input characteristics (and characteristics dependent on them). Calculation of a current distribution by means of these programs is difficult problem. Secondly, such program doesn't allow to know the reasons of distribution change and don't permit to take into account these reasons for using them in the antenna design. Unfortunately, the distribution of a current along the transparent antennas have not been considered in the published papers.

An equation for a current in a transparent antenna [68] is an integral equation with nonzero (impedance) boundary conditions. It is a variant of an equation (1.4), which was considered in the Chapter 1. This equation is valid, if concentrated loads included in the antenna wire and a surface impedance of an antenna are large enough to change a wave propagation constant along the antenna already in the first approximation. In the case of a transparent antenna it is necessary first of all to take into account the surface impedance created by losses in the transparent film. It is equal to $Z = R_{sq1}$. The width b of a transparent antenna can be taken into account afterwards, since this width has smaller effect on the antenna characteristics. The square of a new propagation constant similarly (1.7) is equal to

$$\gamma^2 = k^2 - j\Delta. \tag{12.19}$$

Here $\Delta = 4\pi\omega\varepsilon\chi R$, where $R = R_{sq1}/b$.

As shown in [68], in accordance with (12.19) we obtained that $\gamma = \gamma_0 \exp(-j\varphi)$, $\gamma_0 = \sqrt[4]{k^4 + \Delta^2}$, $\varphi = 0.5 \tan^{-1}(\Delta/k^2)$, from where follows that the current along the antenna is equal to

$$J(z) = J(0)\exp(-\beta z)\frac{\sin \gamma_1 (L - z)}{\sin \gamma_1 L}. \tag{12.20}$$

Here $\gamma_1 = \gamma_0 \cos\varphi$ is the propagation constant and $\beta = \gamma_0 \sin\varphi$ is the decrement (the rate of decrease). In accordance with these expressions the value β at the frequencies 1 and 2 GHz is equal to 11.7 and 13. This means that because of losses in a film the current decreases approximately in three times at a distance $1/\beta$, which is equal to 8.5 and 7.5 cm accordingly. This result leads to an important conclusion: the length of radiating segment depends on β and in the first approximation weakly depends on the frequency. This means that increasing an antenna length does not allow to operate at lower frequencies. These attempts are useless.

In accordance with calculation results, the experimental model of the antenna had a small height (Fig. 12.4, dimensions are given in mm). The radiator is made of CEC005P film. The measurements showed that this model has stable characteristics in the frequency range 2.5–4.5 GHz and higher, close to calculated results. But the exponential decay of current along the antenna leads to a fact that a signal is created by an antenna section located near the power point, and in the rest antenna there is practically no current. Therefore, the input impedance of such an antenna does not have a sharp resonance, and the effective length of the antenna is small in comparison with a monopole manufactured of a metal with high conductivity.

To improve the antenna efficiency and matching it with a cable, between them a metal triangle with a base equal to the radiator width was included. This triangle allows to create an even distribution of current across the entire antenna width that increases the total current and the radiated signal. In order for the current distribution to be uniform and the reflection coefficient to be minimal on the boundary of rectangular and triangular sections and at the cable connection point, the wave impedances of these sections and the cable should be close to each other. Accordingly, the triangular section of the described model, manufactured first as a printed circuit, was replaced by a section of the same CEC005P film (Fig. 12.5).

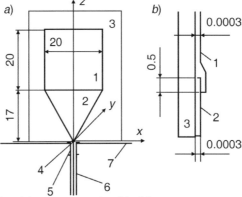

Fig. 12.4: A front view (a) and cross-section (b) of flat transparent antenna: 1—transparent film, 2—metal triangle, 3—glass substrate, 4—soldering point, 5—connector, 6—cable, 7—disc.

As shown in Chapter 6, the wave impedance of a self-complementary antenna depends on the number of flat metal radiators in it and on the connection circuit of these radiators to the cable (to the generator poles). In accordance with Table 6.1 the wave impedance of a symmetric flat antenna with two metal radiators shown in Fig. 6.3a is equal to 17.8π (55.9) Ohm. In the case of an asymmetric antenna (Fig. 6.6) the wave impedance is half as much. But by adjusting the angular width of each radiator, it is possible to obtain the wave impedance equal to 50 Ohm. In Fig. 12.6 reflectivity's of three flat antennas are presented: of antenna with a triangular section, shown in Fig. 7.4 (curve 1), of antenna with one transparent cone (Fig. 12.5a) and arm length equal to $L = 0.045$ m (curve 2) and of antenna with two transparent cones (Fig. 12.5b) and identical arm length (curve 3). As can be seen from Fig. 12.6, an antenna with two cones provides a smooth

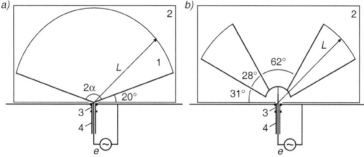

Fig. 12.5: Asymmetrical self-complementary antennas with one (a) and two (b) flat cones: 1—transparent film, 2—glass substrate, 3—connector, 4—cable.

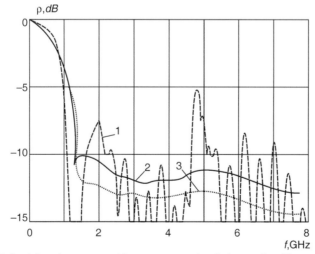

Fig. 12.6: Reflectivity of antennas with transparent triangle (curve 1), with one transparent cone (curve 2) and with two transparent cones (curve 3).

change of reflectivity over a wide frequency range. This result demonstrates a significant advantage of the third option.

12.3 Field of a rectangular loop

One of the extensively used types of antennas is a loop antenna, which can be used, for example, in personal cell phones. For such antennas, a field in the far zone, created by a circular loop, is usually considered. However, the field of a square or rectangular antenna loop is very different from the field of a circular loop. Near fields of a circular loop are analyzed only in a limited number of works [69–72]. About a square loop it is known that if its size is small in comparison with the wavelength, then its field in the far zone in the direction perpendicular to the square side coincides with the field of the circular loop (see, for example, [73]). The advantages of square and rectangular loops make them by attractive and require to execute analysis of their characteristics in the near zone. For example, it is obvious that if the dimensions of the loops are small, but finite in comparison with the wavelength, then in the near zone the shape of the loop will affect the magnitude of the field.

The following results were published in [74].

The scheme of the rectangular loop is presented in Fig. 12.7. The loop of size $a \times b$ lies in a plane xOy, and its center coincides with the origin of coordinates system. It is assumed, that the loop sizes are small $(a,b << \lambda)$, and the amplitude and phase of a current I along a loop wire does not vary. The azimuths of loop tops are $\varphi_{01} = arctg \dfrac{b}{a}$, $\varphi_{02} = \pi - \varphi_{01}$, $\varphi_{03} = \pi + \varphi_{01}$, $\varphi_{04} = -\varphi_{01}$. The azimuth of an observation point P is equal to φ_1. The radius-vector r of a radiation point located on a loop wire depends on an angle φ and is equal to

$$
r = \begin{cases}
\dfrac{a}{2\cos\varphi}, & \varphi_{04} \le \varphi \le \varphi_{01} \\[2mm]
\dfrac{b}{2\sin\varphi}, & \varphi_{01} \le \varphi \le \varphi_{02} \\[2mm]
-\dfrac{a}{2\cos\varphi}, & \varphi_{02} \le \varphi \le \varphi_{03} \\[2mm]
-\dfrac{b}{2\sin\varphi}, & \varphi_{03} \le \varphi \le \varphi_{04}
\end{cases}
\qquad (12.20)
$$

The azimuth component E_φ of loop electrical field is

$$
E_\varphi = -j\omega A_\varphi, \qquad (12.21)
$$

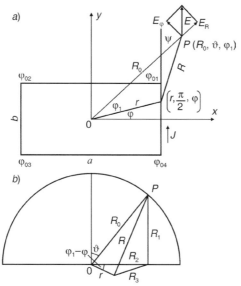

Fig. 12.7: Geometry of the rectangular loop: a—in the loop plane, b—in axonometry.

where $A_\varphi = \dfrac{\mu I}{4\pi} \oint_l \dfrac{e^{-jkR}}{R} \sin\psi dl$ is the appropriate component of a vector potential, μ is permeability, R is a distance from an observation point $(R_0, \vartheta, \varphi_1)$ to an integration point $(r, \pi/2, \varphi)$, ψ is an angle between a current vector and a direction from coordinate origin to an observation point (a straight-line segment R_0), and $k = 2\pi/\lambda$.

In expression (12.21) it is taken into account that the current is closed in a loop, i.e., $div\vec{A} = 0$. As is visible from Fig. 12.7a, the current I at an integration point $(r, \pi/2, \varphi)$ creates at an observation point P not only radial, but also azimuth components of electrical field. Thus, in contrast with a circular loop, on a rectangular loop wire there is no other point (symmetrical concerning a direction from a coordinates origin to an observation point), where current of this point creates at an observation point the radial component with the same magnitude, but opposite sign. It means, that, in addition to component E_φ, the electrical field has also component E_R. Nevertheless, as well as in case of a circular loop, in an antenna far field the field components E_φ and H_ϑ predominate, and

$$H_\vartheta = \frac{1}{\mu} curl_\vartheta \vec{A} = \frac{1}{j\omega\mu R_0} \frac{\partial}{\partial R_0}\left(E_\varphi R_0\right). \qquad (12.22)$$

From Fig. 12.7b it is visible that $R^2 = R_1^2 + R_3^2$, where $R_1 = R_0\cos\vartheta$, $R_3^2 = R_2^2 + r^2 - 2rR_2\cos(\varphi - \varphi_1)$, $R_2 = R_0\sin\vartheta$, i.e.,

$$R^2 = R_0^2 + r^2 - 2rR_0 \sin \vartheta \cos(\varphi - \varphi_1). \tag{12.23}$$

Let

$$R = R_0 \sqrt{1+x}, \tag{12.24}$$

where $x = -\dfrac{2r}{R_0} \sin \vartheta \cos (\varphi - \varphi_1) + \dfrac{r^2}{R_0^2} \ll 1.$

Representing the function $\sqrt{1+x}$ as the sum of an ascending power series and restricting ourselves to the first four terms, we obtain

$$\sqrt{1+x} = 1 + \frac{x}{2} - \frac{x^2}{8} + \frac{x^3}{16} - \cdots$$

Accordingly

$$R = R_0 - r\sin \vartheta \cos(\varphi - \varphi_1) + \frac{r^2}{2R_0} \left[1 - \sin^2 \vartheta \cos^2 (\varphi - \varphi_1) \right] +$$

$$\frac{r^3}{2R_0^2} \sin \vartheta \cos(\varphi - \varphi_1) \left[1 - \sin^2 \vartheta \cos^2 (\varphi - \varphi_1) \right]. \tag{12.25}$$

Using the fact that the difference $R - R_0$ is small, we expand the fraction by the Taylor theorem. In the vicinity of a point $R = R_0$

$$\frac{e^{-jkR}}{R} = \frac{e^{-jkR_0}}{R_0} + (R - R_0) \frac{\partial}{\partial R_0} \left(\frac{e^{-jkR_0}}{R_0} \right) + \frac{1}{2}(R - R_0)^2 \frac{\partial^2}{\partial R_0^2} \left(\frac{e^{-jkR_0}}{R_0} \right) + \frac{1}{6}(R - R_0)^3 \frac{\partial^3}{\partial R_0^3} \left(\frac{e^{-jkR_0}}{R_0} \right) + \cdots \tag{12.26}$$

The substitution of derivatives in (12.26) and reducing similar terms gives

$$f(\varphi) = \frac{e^{-jk(R-R_0)}}{R/R_0} = f_1(\varphi) + jk(1+\alpha)\sin \vartheta \left[f_2(\varphi)\cos \varphi_1 + f_3(\varphi)\sin \varphi_1 \right] +$$

$$\frac{k^2}{2} \left[\alpha(1+\alpha) - \frac{1}{2}(1+3\alpha+3\alpha^2)\sin^2 \vartheta \right] f_4(\varphi) -$$

$$-\frac{k^2}{2}(1+3\alpha+3\alpha^2)\sin^2 \vartheta \left[f_5(\varphi)\cos 2\varphi_1 + f_6(\varphi)\sin 2\varphi_1 \right] +$$

$$j\frac{k^3\alpha}{2}(1+3\alpha+3\alpha^2)\sin \vartheta \left[f_7(\varphi)\cos \varphi_1 + f_8(\varphi)\sin \varphi_1 \right] -$$

$$-j\frac{k^3}{8} \left(\frac{1}{3}+2\alpha+5\alpha^2+5\alpha^3 \right)\sin^3 \vartheta *$$

$$\left[f_9(\varphi)\cos 3\varphi_1 + f_{10}(\varphi)\sin 3\varphi_1 + 3f_7(\varphi)\cos \varphi_1 + 3f_8(\varphi)\sin \varphi_1 \right], \tag{12.27}$$

where

$$f_1(\varphi)=1, f_2(\varphi)=r\cos\varphi, f_3(\varphi)=r\sin\varphi, f_4(\varphi)=r^2, f_5(\varphi)=r^2\cos2\varphi, f_6(\varphi)=r^2\sin2\varphi,$$
$$f_7(\varphi)=r^3\cos\varphi, f_8(\varphi)=r^3\sin\varphi, f_9(\varphi)=r^3\left(4\cos^3\varphi-3\cos\varphi\right), f_{10}(\varphi)=r^3\left(3\sin\varphi-4\sin^3\varphi\right).$$

In expression (12.27) the symbol $a=1/(jkR_0)$. It is easy also to show, that on the loop sides

$$\sin\psi=\begin{cases}\cos\varphi_1\\\sin\varphi_1\\-\cos\varphi_1\\-\sin\varphi_1\end{cases} \qquad dl=\begin{cases}\dfrac{rd\varphi}{\cos\varphi}=\dfrac{ad\varphi}{2\cos^2\varphi} & \varphi_{04}\le\varphi\le\varphi_{01}\\[2mm]\dfrac{bd\varphi}{2\sin^2\varphi} & \varphi_{01}\le\varphi\le\varphi_{02}\\[2mm]\dfrac{ad\varphi}{2\cos^2\varphi} & \varphi_{02}\le\varphi\le\varphi_{03}\\[2mm]\dfrac{bd\varphi}{2\sin^2\varphi} & \varphi_{03}\le\varphi\le\varphi_{04}\end{cases} \qquad (12.28)$$

Substituting (12.20), (12.27) and (12.28) in expression for A_φ, results in:

$$A_\varphi=\frac{\mu I}{4\pi}\frac{e^{-jkR_0}}{R_0}\oint_\varphi f(\varphi)\sin\psi dl=\frac{\mu I}{4\pi}\frac{e^{-jkR_0}}{R_0}jkSF(f)\sin\vartheta, \quad (12.29)$$

where

$$jkSF(f)\sin\vartheta=\frac{a}{2}\cos\varphi_1\left(\int_{\varphi_{04}}^{\varphi_{01}}-\int_{\varphi_{02}}^{\varphi_{03}}\right)\frac{f(\varphi)d\varphi}{\cos^2\varphi}+\frac{b}{2}\sin\varphi_1\left(\int_{\varphi_{01}}^{\varphi_{02}}-\int_{\varphi_{03}}^{\varphi_{04}}\right)\frac{f(\varphi)d\varphi}{\sin^2\varphi},$$

$$(12.30)$$

and $S=ab$ is loop area. Designate

$$F_i=jkSF(f_i)\sin\vartheta. \qquad (12.31)$$

Then from (12.27)

$$jkSF(f)\sin\vartheta=F_1+jk(1+\alpha)\sin\vartheta(F_2\cos\varphi_1+F_3\sin\varphi_1)+$$
$$\frac{k^2}{2}\left[\alpha(1+\alpha)-\frac{1}{2}(1+3\alpha+3\alpha^2)\sin^2\vartheta\right]F_4-\frac{k^2}{2}(1+3\alpha+3\alpha^2)\sin^2\vartheta*$$
$$(F_5\cos2\varphi_1+F_6\sin2\varphi_1)+j\frac{k^3\alpha}{2}(1+3\alpha+3\alpha^2)\sin\vartheta(F_7\cos\varphi_1+F_8\sin\varphi_1)-$$
$$-j\frac{k^3}{8}\left(\frac{1}{3}+2\alpha+5\alpha^2+5\alpha^3\right)\sin^3\vartheta(F_9\cos3\varphi_1+F_{10}\sin3\varphi_1+3F_7\cos\varphi_1+3F_8\sin\varphi_1).$$

The integration of expressions (12.31) gives

$$F_1 = F_4 = F_5 = F_6 = 0, \quad F_2 = ab\cos\varphi_1, \quad F_3 = ab\sin\varphi_1, \quad F_7 = \frac{a^3 b}{4}\left(1 + \frac{b^2}{3a^2}\right)\cos\varphi_1,$$

$$F_8 = \frac{ab^3}{4}\left(1 + \frac{a^2}{3b^2}\right)\sin\varphi_1, \quad F_9 = \frac{a^3 b}{4}\left(1 - \frac{b^2}{a^2}\right)\cos\varphi_1, \quad F_{10} = -\frac{a^3 b}{4}\left(1 - \frac{a^2}{b^2}\right)\sin\varphi_1.$$

And accordingly

$$F(f) = 1 + \alpha + \frac{k^2}{8}\left[\left(a^2 + \frac{b^2}{3}\right)\cos^2\varphi_1 + \left(b^2 + \frac{a^2}{3}\right)\sin^2\varphi_1\right]\alpha\left(1 + 3\alpha + 3\alpha^2\right) -$$

$$- \frac{k^2}{8}\left(a^2\cos^2\varphi_1 + b^2\sin^2\varphi_1\right)\left(\frac{1}{3} + 2\alpha + 5\alpha^2 + 5\alpha^3\right)\sin^2\vartheta. \tag{12.32}$$

The magnitude A_φ is found from (10); the magnitude E_φ is equal to

$$E_\varphi = 30k^2 IS \frac{e^{-jkR_0}}{R_0} F(f)\sin\vartheta. \tag{12.33}$$

If the loop sizes are indefinitely small in comparison with a wavelength, then

$$E_\varphi = 30k^2 IS(1 + \alpha)\frac{e^{-jkR_0}}{R_0}\sin\vartheta. \tag{12.34}$$

This expression coincides with the known expression for the azimuth field component of the elementary circular loop which radius is small in comparison with a wavelength (see [75]). In it the circular loop area is replaced by the rectangular loop area.

Comparing the formulas (12.33) and (12.34), it is easy to see, that in (12.34) F(f) = 1 + α, i.e., from expression (12.32) only the first two terms are taken. The remaining terms are proportional to the squares of the ratios of loop linear size to a wavelength and higher powers of α (to the inverse powers of distance to an observation point). From (12.32) the field magnitude in a near field depends on the observation point azimuth. The rectangular (noncircular) loop shape causes the radial field component

$$E_R = -j\omega A_R, \tag{12.35}$$

where $A_R = \dfrac{\mu I}{4\pi}\displaystyle\oint_l \dfrac{e^{-jkR}}{R}\cos\psi dl$ is the vector potential radial component, and

$$\cos \psi = \begin{cases} \sin \varphi_1 & \varphi_{04} \le \varphi \le \varphi_{01} \\ -\cos \varphi_1 & \varphi_{01} < \varphi \le \varphi_{02} \\ -\sin \varphi_1 & \varphi_{02} \le \varphi \le \varphi_{03} \\ \cos \varphi_1 & \varphi_{03} \le \varphi \le \varphi_{04}. \end{cases} \quad (12.36)$$

Subject to (12.36) the expression for A_R assumes the form

$$A_R = \frac{\mu I}{4\pi} \frac{e^{-jkR_0}}{R_0} jkSG(f)\sin \vartheta, \quad (12.37)$$

where

$$jkSG(f)\sin \vartheta = \frac{a}{2}\sin \varphi_1 \left(\int_{\varphi_{04}}^{\varphi_{01}} - \int_{\varphi_{02}}^{\varphi_{03}} \right) \frac{f(\varphi)d\varphi}{\cos^2 \varphi} - \frac{b}{2}\sin \varphi_1 \left(\int_{\varphi_{01}}^{\varphi_{02}} - \int_{\varphi_{03}}^{\varphi_{04}} \right) \frac{f(\varphi)d\varphi}{\sin^2 \varphi}, \quad (12.38)$$

and the function $f(\varphi)$ is consistent with the formula (12.37). Designate

$$G_i = jkSG(f_i) \sin \vartheta, \quad (12.39)$$

then the expression for $jkSG(f) \sin \vartheta$ is similar to the expression for $jkSF(f) \sin \vartheta$ with change of quantity F_i to G_i. The integration of equation (12.39) gives

$$G_1 = G_4 = G_5 = G_6 = 0, \quad G_2 = ab\sin\varphi_1, \quad G_3 = -ab\cos\varphi_1, \quad G_7 = \frac{a^3b}{4}\left(1 + \frac{b^2}{3a^2}\right)\sin\varphi_1,$$

$$G_8 = \frac{ab^3}{4}\left(1 + \frac{a^2}{3b^2}\right)\cos\varphi_1, \quad G_9 = \frac{a^3b}{4}\left(1 - \frac{b^2}{a^2}\right)\sin\varphi_1, \quad G_{10} = -\frac{a^3b}{4}\left(1 - \frac{b^2}{a^2}\right)\cos\varphi_1,$$

and accordingly

$$G(f) = \frac{k^2}{4}\sin 2\varphi_1 \left[\alpha\left(\frac{1}{3} + \alpha + \alpha^2\right)(a^2 + b^2) - \frac{1}{8}\left(\frac{1}{3} + 2\alpha + 5\alpha^2 + 5\alpha^3\right)(a^2 + 3b^2)\sin^2 \vartheta \right]. \quad (12.40)$$

The magnitude A_R is found from (18); the magnitude E_R is equal to

$$E_R = 30k^2 IS \frac{e^{-jkR_0}}{R_0} G(f)\sin \vartheta. \quad (12.41)$$

As seen from (12.41), the radial field component contains only terms which are proportional to the squares of the ratios of loop linear size to a wavelength, i.e., it is small in comparison to the azimuth component. In addition, the radial component substantially depends on the azimuth. This component vanishes in the directions which are perpendicular to the rectangle sides, because in this case each loop conductor point opposes, creating the field dE_R at the observation point, the point symmetric to that perpendicular, where the current of this point will create the field $-dE_R$ at the observation point. Peaks of E_R occur at angle $\pi/4$ with respect to the given directions.

In Fig. 12.8 the calculated results for a field E_ψ are indicated. These results are depending on a distance R_0/λ between the rectangular loop center and an observation point (in wavelengths). The loop dimensions are assumed equal to $a = 2b = 1/(2k)$, the observation point azimuth is $\varphi = \pi/2$, and angle ϑ varies from $\pi/6$ to $\pi/2$. The field magnitude is referred to its significance for $\vartheta = \pi/2$, $R_0/\lambda = 0.1$.

In Fig. 12.9 the curves for E_φ of the similarly shaped loops ($a = 2b$), but with different areas (in wavelengths) are presented, and $\vartheta = \pi/2$, $\varphi_1 = \pi/2$. The fields of different area loops are the same in the far field ($IS = const$), and they are referred to the field of a loop with $a = 1/(2k)$ and $R_0/\lambda = 0.1$. The curve 1 was calculated under the approximate formula (15) that is for the loop, which dimensions are indefinitely small in comparison with a wavelength. It could be seen that the loop field in the near field calculated under the exact formula essentially differs from the approximate one.

In Fig. 12.10 similar curves for various aspect ratios are given. The fields of different loops are the same in the far field ($IS = const$), and they are referred to the field of a loop with $b = a/2 = 1/(4k)$ and $R_0/\lambda = 0.1$.

In Fig. 12.11 the results for the radial field component E_R are presented. The loop dimensions are $a = 2b$, the observation point azimuth is $\varphi_1 = \pi/4$, and the angle ϑ is $\vartheta = \pi/2$. The magnitudes E_R of the loops with the same fields E_φ in the far zone ($IS = const$) are referred to the magnitude E_φ of the loop with $a = 1/(2k)$ and $R_0/\lambda = 0.1$. As may be seen, E_R is small in comparison to E_φ, and is equal to zero in case of an elementary loop, which dimensions are indefinitely small in comparison with a wavelength.

The azimuth and radial field components of a rectangular loop in far and near field are calculated. The results are compared to the field of an elementary circular loop. It is demonstrated that loop dimensions and shape substantially affect its field magnitude, especially in the near field.

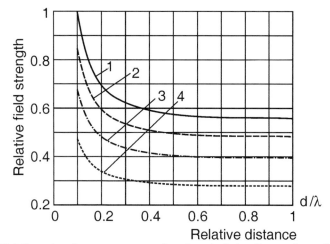

Fig. 12.8: Relative azimuth components as a function of relative distance for various values of angle ϑ. Graphs 1–4: $\vartheta = \pi/2, \pi/3, \pi/4, \pi/6$.

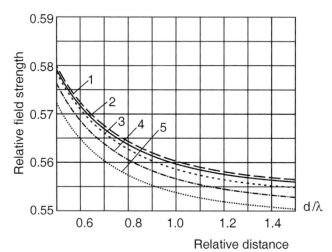

Fig. 12.9: Relative azimuth components for similar form and different sizes loops and the same fields in the far zone. Graphs 1–5: $a = 0, \lambda/8\pi, \lambda/4\pi, 3\lambda/8\pi, \lambda/2\pi$.

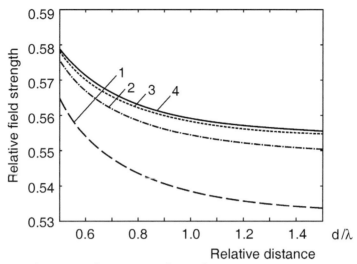

Fig. 12.10: Relative azimuth components for similar size *a*, various aspect ratios and the same fields in the far zone. Graphs 1–4: $b = 2a, a, a/2, a/4$.

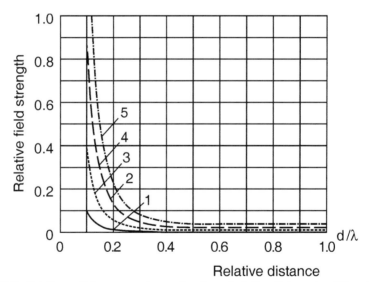

Fig. 12.11: Relative radial components as a function of relative distance for different sizes loops and the same fields in the far zone. Graphs 1–5: $a = 0, \lambda/8\pi, \lambda/4\pi, 3\lambda/8\pi, \lambda/2\pi$.

References

1. Miller, M.A. 1954. Application of uniform boundary conditions in the theory of thin antennas. Journal of Technical Physics 24: 1483–1495 (in Russian).
2. Leontovich, M.A. and Levin, M.L. 1944. On the theory of oscillations excitation in the linear radiators. Journal of Technical Physics 14: 481–506 (in Russian).
3. Vainshtein, L.A. 1957. Electromagnetic Waves. Sovetskoye Radio, Moscow. USSR (in Russian).
4. Glushkovsky, E.A., Levin, B.M. and Rabinovich, E.Ya. 1967. Integral equation for the current in thin impedance radiator. Radiotechnics 22: 12, 18–23 (in Russian).
5. Glushkovsky, E.A., Izrailit, A.B., Levin, B.M. and Rabinovich, E.Ya. 1968. Methods of calculating linear impedance radiators. Radiotechnics 23: 1, 40–46 (in Russian).
6. Glushkovsky, E.A., Izrailit, A.B., Levin, B.M. and Rabinovich, E.Ya. 1967. Linear antennas with surface impedance changing along its length. Antennas 2: 154–165 (in Russian).
7. Finck, L.M. 1984. Signals, Interferences, Mistakes ... Notes on Some Surprises, Paradoxes and Errors in the Theory of Communication. Radio and Communication, Moscow. USSR (in Russian).
8. Popovic, B.D. 1973. Theory of cylindrical antennas with lumped impedance loadings. The Radio and Electronic Engineer 43: 3, 243–248.
9. Levin, B.M. 1998. Monopole and Dipole Antennas for Marine-Vehicle Radio Communications. Абрис, S.-Petersburg. Russia (in Russian).
10. Wu, T.T. and King, R.W.P. 1965. The cylindrical antenna with non-reflecting resistive loading. IEEE Transactions AP-13: 3, 369–373.
11. Rao, B.L.J., Ferris, J.E. and Zimmerman, W.E. 1969. Broadband characteristics of cylindrical antennas with exponentially tapered capacitive loading. IEEE Transactions AP-17: 2, 145–151.
12. Popovic, B.D. and Dragovic, M.B. 1972. Cylindrical antennas with constant capacitive loadings. Electronics Letters 8: 15, 396–398.
13. Kanda, M. 1980. The characteristics of a linear antenna with tapered resistive and capacitive loading. Proceedings of Jntern. Symp. Digest Antennas and Propagation, Quebec. Canada 2: 696–699.
14. Dragovic, M.B. and Popovic, B.D. 1981. Limits of VSWR for optimal broadband capacitively loaded cylindrical antennas versus their length. Proceedings of the Second Jntern. Conf. on Antennas and Propagation, Heslington 1: 343–347.
15. Levin, B.M. and Yakovlev, A.D. 1985. Antenna with loads as impedance radiator with impedance changing along its length. Soviet Journal of Communications Technology and Electronics 30: 1, 25–33.
16. Levin, B.M. 1990. Use of loads for creating a given current distribution along a dipole. Soviet Journal of Communications Technology and Electronics 35: 8, 1581–1589.
17. Himmelblau, D. 1972. Applied Nonlinear Programming. McGraw-Hill, New York. USA.

18. Levin, B.M., Fradin, A.Z. and Yakovlev, A.D. 1988. Antennas with capacitive loads and decrease of reradiators influence. Proceedings of the 9th Intern. Wroclaw Symposium on Electromagnetic Compatibility, Wroclaw. Poland 1: 327–332.

19. Levin, B.M. 1993. V-dipole with capacitive loads. Soviet Journal of Communications Technology and Electronics 38: 3, 400–408.

20. Levin, B.M. 2016. Directivity of thin antennas. Proceedings of the 21th Intern. Seminar/ Workshop on Direct and Inverse Problems of Electromagnetic and Acoustic Wave Theory DIPED, Tbilisi. Georgia.

21. Pistolkors, A.A. 1944. General theory of diffraction antennas. Journal of Technical Physics 14: 12, 693–701 (*in Russian*).

22. Pistolkors, A.A. 1948. Theory of the circular diffraction antenna. Proceedings of IRE 1: 56–60.

23. Neyman, L.R. and Kalantarov, P.L. 1959. Theoretical Background of Electro Engineering, part 3. Moscow—Leningrad. Russia (*in Russian*).

24. Iossel, Yu.Ya., Kochanov, E.S. and Strunsky, M.G. 1981. Calculation of Electrical Capacitance. Energoisdat, Leningrad. Russia (*in Russian*).

25. Chaplin, A.F., Buchazky, M.D. and Mihailov, M.Yu. 1983. Optimization of director-type antennas. Radiotechnics: 7, 79–82 (*in Russian*).

26. Carrel, R.L. 1961. The design of log-periodic dipole antennas. IRE Intern. Convention Record, part 1: 61–75.

27. Yakovlev, A.F. and Pyatnenkov, A.E. 2007. Wide-Band Directional Antennas Arrays from Dipoles. S.-Petersburg. Russia (*in Russian*).

28. Booker, H.G. 1946. Slot aerials and their relation to complementary wire aerials (Babinet's principle). The Journal of the Institution of Electrical Engineers, Part IIIA 4: 620–626.

29. Mushiake, Y. 1996. Self-Complementary Antennas: Principle of Self-Complementarity for Constant Impedance. Springer, London. UK.

30. Rumsey, V.H. 1957. Frequency independent antennas. 1957 IRE National Convention Record, Part 1: 114–118.

31. Rumsey, V.H. 1966. Frequency Independent Antennas. Academic Press. USA.

32. Levin, B.M. and Markov, V.G. 1997. Method of Complex Potential and Antennas. Ship Electrical Engineering and Communication, St.-Petersburg. Russia (*in Russian*).

33. Carrel, R.L. 1958. The characteristic impedance of two infinite cones of arbitrary cross-section. IEEE Transactions AP-6: 2, 197–201.

34. Buchholz, H. 1957. Elektrische und Magnetische Potentialfelder. Berlin. Germany.

35. Levin, B.M. 2013. The Theory of Thin Antennas and Its Use in Antenna Engineering, Bentham Science Publishers.

36. Korn, G. and Korn, T. 1961. Mathematical Handbook for Scientists and Engineers. McGraw-Hill, New York, Toronto, London.

37. Belousov, S.P., Gurevich, R.V., Kliger, G.A. and Kuznetsov, B.D. 1979. Antennas for Broadcasting and Radio Communication, part 1. Shortwave Antennas. Sviyaz, Moscow, Russia (*in Russian*).

38. Pistolkors, A.A. 1947. Antennas. Sviazizdat, Moscow. USSR (*in Russian*).

39. Vershkov, M.V., Levin, B.M. and Fraiman, S.S. 1972. Antenna with meandering load. Proceedings of CNII MF 151: 73–80 (*in Russian*).

40. Levin, B.M. 1976. Meandering load with arbitrary number of wires. Proceedings of CNII MF 216: 130–139 (*in Russian*).

41. Kusnezov, V.D. 1955. Shunt radiators. Radiotechnics 10: 57–65 (*in Russian*).

42. Levin, B.M. 1976. Impedance folded radiator. Antennas 23: 80–90 (*in Russian*).

43. Leontovich, M.A. 1945. Theory of forced electromagnetic oscillations in thin conductors of arbitrary cross section and its applications to calculation of some antennas. Proceedings of NII MPSS 1: 1 (*in Russian*).

44. Toftgard, J. and Hornslets, S.N. 1993. Effects on portable antennas of the presence of a person. IEEE Transactions AP-41: 6, 739–746.

45. Jensen, M.A. and Rahmat-Samii, Y. 1995. EM Interaction of handset antennas and a human in personal communications. Proceedings of the IEEE 81: 1, 7–17.
46. Bank, M. and Levin, B. 2007. The development of the cellular phone antenna with a small radiation of human organism tissues. IEEE Antennas Propagation Magazine 4: 65–73.
47. Levin, B. and Kondratiev, M. 2017. An antenna for a cell phone. Proceedings of the 11th European Conference on Antennas and Propagation EuCAP, Paris. France.
48. King, R.W.P. 1956. Theory of Linear Antennas. Harvard University Press, Cambridge. USA.
49. Belousov, S.P. and Kliger, G.A. 1982. Analysis of wire radiators. Proceedings of NII Radio 3: 5–11.
50. Popovic, B.D. and Surutka, J.V. 1972. Cylindrical cage antenna. Nachrichtentechn. Fachber. 45: 51.
51. Pantic, Z.Z. 1979. Cage-dipole antenna driven by a two-wire line. Archiv fur Electronik and Ubertragungs Technics 33: 7/8, 329–330.
52. Vershkov, M.V. 1979. Ship antennas. Shipbuilding, Leningrad, USSR (*in Russian*).
53. Levin, B.M. and Mirotvorsky, O.B. 1986. Multi-radiator antenna with a resistor. Antennas 33: 94–100 (*in Russian*).
54. Levin, B.M. and Yakovlev, A.F. 1992. About one method of widening an antenna operation range. Radiotechnics and Electronics Engineering 1: 55–64 (*in Russian*).
55. DuHamel, R.P. and Isbell, D.E. 1957. Broadband logarithmically-periodic antennas structures. IRE National Convention Record, part 1: 119–128.
56. DuHamel, R.P. and Ore, F.R. 1958. Logarithmically-periodic antenna designs. IRE National Convention Record, part 1: 139–151.
57. DuHamel, R.P. and Berry, D.C. 1958. Arrays of log-periodic antennas. IRE Wescon. Convention Record, part 1: 161–174.
58. Isbell, D.E. 1960. Log-periodic dipole arrays. IRE Transactions AP-3: 260–267.
59. Yazkevich, V.A. and Lapizky, V.M. 1979. Exact and approximate methods of calculating log-periodic antennas. Izvestiya vusov of USSR – Radioelektronika 5: 69–72 (*in Russian*).
60. De Vito, G. and Strassa, G.B. 1973. Comments on the design of log-periodic dipole antennas. IEEE Transaction AP-3: 303–309.
61. Yakovlev, A.F. and Pyatnenkov, A.E. 2007. Wide-Band Directional Antennas Arrays From Dipoles. S.-Petersburg. Russia (*in Russian*).
62. Shifrin, Y.S. and Luchaninov, A.I. 2003. Microwave devices with the distributed nonlinearity. Proceedings of Intern. Conf. on Antenna Theory and Techniques, Sevastopol. Ukraine: 81–86.
63. Guan, N. et al. 2008. Antennas made of transparent conductive films. PIERS Online 4: 1, 116–120.
64. Nesterenko, M.V. 2010. Analytical methods in the theory of thin impedance vibrators. Progress in Electromagnetics Research B 21: 299–328.
65. Peter, T. et al. 2014. A novel transparent UWB antenna for photovoltaic solar panel integration and RF energy harvesting. IEEE Transactions AP-62: 4, 1844–1853.
66. Hautcoeur, J., Colombel, F., Castel, X., Himdi, M. and Motta Cruz, E. 2009. Optically transparent monopole antenna with high radiation efficiency manufactured with silver grid layer (AgGL). Electronics Letters 45: 20, 1014–1016.
67. Saberin, J.R. and Furse, C. 2012. Challenges with optically transparent patch antennas. IEEE Transactions AP-54: 3, 10–16.
68. Levin, B.M. 2016. Flat transparent antenna. Radio Science Bulletin: 356, 32–40.
69. Green, F.M. 1967. The near-zone magnetic field of a small circular-loop antenna. J. Res. NSB, 71-C: 4, 319–326.
70. Werner, D.H. 1996. An exact integration procedure for vector potentials of thin circular loop antennas. IEEE Transactions AP-44: 2, 157–165.
71. Overfelt, P.L. 1996. Near fields of the constant current thin circular loop antenna of arbitrary radius. IEEE Transactions AP-44: 2, 166–171.

72. Li, L.W., Leong, M.S., Kooi, P.S. and Yeo, T.S. 1997. Exact solutions of electromagnetic fields in both near and far zones radiated by thin circular-loop antennas: a general representation. IEEE Transactions AP-45: 12, 1741–1748.
73. Kraus, J.D. 1988. Antennas. McGraw-Hill, Boston.
74. Levin, B.M. 2004. Field of a rectangular loop. IEEE Transactions AP-52: 4, 948–952.
75. Balanis, C.A. 1982. Antenna Theory: Analysis and Design. Harper and Row, New York.

Index